A BIRDWATCHERS' GUIDE TO TURKEY

IAN GREEN AND
NIGEL MOORHOUSE

Illustrations by Mike Langman

Prion Ltd.
Perry

ACKNOWLEDGEMENTS

We would like to thank the Ornithological Society of the Middle East, and in particular Richard Porter and Rodney Martins for all their help and advice over past trips to Turkey. We are grateful for all the trip reports and comments supplied by various people over the past few years, especially those from Magnus Ullman, Jan Vermeulen and Richard Webb. We would like to thank all those who have travelled around Turkey with us, especially Günther Bauer, Julia Bustin, Fiona Dunbar, Steve Moorhouse, Owen Mountford, John Overfield and Martin Robson, and all those who have joined us on tours to Turkey. We would also like to thank the following for their help and friendliness in enriching our knowledge of the area; Mehmet Acar, Paulo Gelati, June Haimoff, Max Kasparek, Sven Larsson, Ali Safak, Levent Sencan, Levent Yasar, and the French teacher in Birecik, who's name we can't remember, but who, we're sure, can remember his trip up the wadis with us! Special thanks to Susannah and Fiona for putting up with years of talk about 'the book'. Lastly we should like to thank the people of Turkey for making all our visits so pleasurable and enlivening.

This book is dedicated to Ian's parents Val and Tony Green, and the memory of Nigel's grandmother, Mrs E Wilkinson.

Nigel Moorhouse was born in 1964 in Preston. He began birdwatching at an early age and graduated in Zoology at Leeds University in 1985. He has travelled extensively and has worked as a birdwatching tour leader. He is currently studying the ecology of White-headed Duck in Turkey.

Ian Green was born in Sussex in 1963 but spent his formative years on a farm in Devon where he developed a keen interest in all forms of natural history. After graduating in ecology at the University of East Anglia he has worked as a wildlife tour guide and now runs Greentours, his own natural history holiday business. An expedition to study the ornithological importance of wetlands in eastern and central Turkey was Ian's introduction to a country which has drawn him back again and again.

CONTENTS

INTRODUCTION

Turkey is composed of a wide diversity of habitats, with mountains, deserts, coasts, forests and lakes throughout a variety of climatic zones. This accounts for its wealth of birds and it has more species of breeding bird than any other country in the Western Palearctic. There are tremendous opportunities for the visiting naturalist, with a huge variety of plants, amphibians, reptiles and mammals complementing the birdlife.

Turkey has been an important political area for many years. Modern civilisation developed in the Fertile Crescent between the Tigris and Euphrates rivers, and since then Turkey has been occupied by a wide variety of peoples. There are a great many books on the various phases of Anatolia's amazingly complex history, and a visit to a local library, or bookshop, should provide ample reading material. While travelling through Turkey one cannot help but notice the plethora of ancient, and not so ancient ruins, each one coming from a different people or period; from the Hittites (Bronze Age), through the Phrygians, Lycians, Urartians, Cyrus of Persia, Alexander the Great's rapid conquest, and almost as rapid decline, the Romans, and eventually to the first invasion of Turkic peoples and the formation of the Seljuk Empire. The latter, and their immediate conquerors the Ottomans, were, in their day, two of the most important and influential empires in the world, leaving some very impressive architecture. The zenith of the Ottoman Empire was under Süleyman the Magnificent (1520-1566), whereafter there was a prolonged and steady decline culminating in the defeat by the Allies in the First World War, after the Commitee of Union and Progress (ruling even though there was still a Sultan on the throne) had entered the war on the German side. The Allies intended to dismember Anatolia as rewards for various alliances, and in encouraging the Greeks to take Smyrna (İzmir) in 1919 they precipitated the Turkish War of Independance which lasted from 1920 until 1922. The prime mover on the Turkish side was a general named Mustafa Kemal, who eventually brought them to victory, and not surprisingly became a national hero. He declared Turkey a Republic and undertook a number of quite astounding reforms. In 1924 Turkey adopted a constitution and banned polygamy and the fez, followed by the adoption of Western style laws and the institution of civil marriage. Turkey became a secular state (i.e. there is no state religion, even though over 95% of the population is Muslim) and perhaps most difficult of all; the Arabic alphabet was replaced with a modified Latin version and all 'civil servants' were required to learn and use the new alphabet within three years. Kemal took the name of Atatürk (father Turk) and presided over the formation of today's Turkey. He attained a near god-like status and although he died in 1938, he is still a very important figure for Turks today, and statues of him can be seen in nearly all settlements bigger than a small village. Since his death Turkey has managed to continue, if somewhat haltingly, with Atatürk's ideas of progress along Western lines, but with Islamic Turkey's own individuality. The democratic process has, on a number of occasions in the last thirty years, failed, and required the intervention of the military.

The future looks very interesting for this large and diverse country. Its importance to NATO has declined with the collapse of the old Soviet regime, and so, while still looking westwards for economic links, Turkey has found itself in a position of great influence, not only in the Middle East, but also with the many Turkish speaking states which have emerged to the north and east, indeed all the way to China.

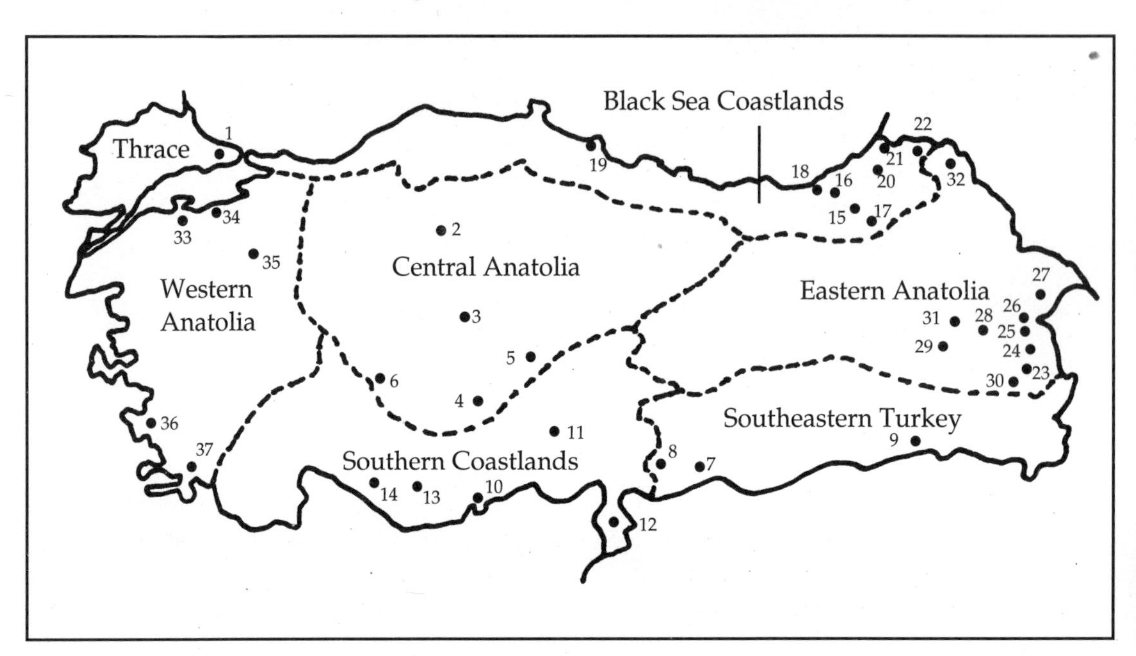

1 İstanbul
2 Soğuksu National Park
3 Kulu Gölü
4 Ereğli Marshes
5 Sultan Marshes
6 The Lake District
7 Birecik and Halfeti
8 Yeşilce
9 İdil and Cizre
10 Göksu Delta
11 Demirkazık
12 Nur Dağları
13 Sertavul
14 Akseki
15 Sivrikaya
16 Uzüngöl
17 İspir
18 Sumela Monastery
19 Kızılırmak Delta
20 Kaçkar Dağları
21 Borçka
22 Çam Geçidi (Şavşat)
23 Van
24 Erçek Gölü
25 Bendimahi Marsh
26 Şelale Waterfall (Gönderme) and Çaldıran Ovası
27 Dogubeyazıt
28 Arin Golu and Suphan Dağı
29 Nemrut Dağı
30 South Van Gölü
31 Bulanık
32 Ardahan Ovası and Çıldır Gölü
33 Manyas Gölü
34 Koçaçay Delta
35 Uludağ
36 Bafa Gölü
37 Dalyan

PRE-TOUR INFORMATION

Visas and passports

A valid passport is required to enter Turkey. Most Europeans do not need a visa, but British nationals do now require one. This costs £10 payable on entry in Sterling. No inoculations are required to gain entry.

Currency and Exchange Rate

The Turkish currency is the Turkish Lira (TL), and the exchange rate is highly variable due to a high inflation rate. There are export and import limits but it isn't worth taking any currency out of the country since it is not accepted anywhere else.

Field Guides

Useful field guides include 'Birds of Britain and Europe with the Middle East and North Africa' by Heinzel, Fitter and Parslow, 'Birds of the Middle East and North Africa' by Hollom, Porter, Christensen and Willis, and 'Birds of Europe, with North Africa and the Middle East' by Lars Jonsson.

Photography

Be careful when using binoculars, telescopes and cameras in sensitive areas, such as military installations, presidential palaces, railway lines etc - these are usually well marked. Film is available in most towns, but expensive, so take as much as you need with you.

Emergencies

Consular addresses in Turkey are:

British Embassy, Şehit Ersan Caddesi 46/A, Çankaya, Ankara. Tel: (4) 427 4310.

British Consulate, Meşrutiyet Caddesi 34, Tepebaşı, Beyoğlu, İstanbul. Tel: (1) 244 7540/5

British Vice-Consulate, Mahmut Eşat Bozkurt Caddesi, 1442 Sokuk No 49, İzmir. Tel: (51) 63 51 51

TRAVEL INFORMATION

International Airports

Turkey is well served with airports. International flights arrive at İstanbul (or more rarely, Ankara), and charters regularly serve the Mediterranean and Aegean holiday areas of Antalya, Dalaman and İzmir.

By road

The easiest and most rewarding way to travel round Turkey is by car. Driving there from Europe will take at least three days from France (two is possible but not recommended), and is an interesting way of getting there if you have unlimited time. If you do take your own vehicle you should obtain a green card with an insurance company that covers the Asian as well as the European part of Turkey.

Local airports

Internal flights are inexpensive and serve most cities (most of these go via Ankara or İstanbul). The airports are often some distance from the cities which they serve, but buses are available at most and taxis at all airports.

Car hire

When hiring a car it is best to book it through an international agency before setting off, since hiring a car from within the country can be difficult (due to the lack of cars available) and actually more expensive. For people short of time but wanting to see most of the country, it is a good idea to catch an internal flight to the furthest point of call and have a hire car waiting there for a one-way rental. One-way rental is available from most agencies and airports and generally only a little more expensive. One recommended method to try to see all the specialities in two or three weeks is to catch an internal flight to Trabzon, and pick up a car there, returning it to your departure airport.

The main roads in Turkey are generally good, although some have bad potholes, so care is required. Occasionally a main road may become a cobbled or stone track for a short distance, especially in mountainous areas. Off the main roads, there are dirt tracks of varying standard, most of which are drivable. Roads can become blocked, especially in winter, and landslips are not uncommon in spring along the Black Sea coast. Fortunately, roads are cleared fairly quickly, but expect the odd detour off the main roads.

Driving in Turkey is not for the faint hearted. In towns and villages, beware of cycles, domestic animals and people stepping out into the road without looking (don't believe that they won't do it, they will). In İstanbul, and to a certain extent Ankara, don't expect to get anywhere fast, and be prepared to be forceful when changing lanes etc. In the country, there is generally very little traffic, except for the main oil tanker routes to Syria and Iraq. The road across the Nur Dağları east of Adana is packed with trucks and the passes will have them overtaking up to three abreast, with none travelling more than 20 mph (33 kph), so be very careful here and don't take any risks.

Petrol stations are frequent on all major routes, and some lesser ones. If in doubt, fill up as often as possible. Some petrol stations have restaurants, and many have tyre repair shops. Mechanics can be found in most towns, and in the larger ones they are often found in one area - the Oto Sanayii. Turkish mechanics can repair virtually any car, mainly by cobbling something up. Turks generally use Renaults (Murat), Fiats (Tofaş) and Fords (Otosan). Don't be surprised if you are fed and offered tea in a garage. When leaving the car unattended, take the precaution of locking it up and keeping valuables out of sight.

Buses, Coaches and Trains

Buses and coaches are cheap and fast and go everywhere, and this is how most Turks travel (very few have private cars). Coaches ply the routes between all major towns and cities, and it is advisable to book in advance if possible. This can be done at one of several booking offices in town or at the bus station (Otogar). Remember to check the departure point. The coach's destination is usually displayed, and quite commonly painted on the coach itself. They are very comfortable and have numerous tea breaks along the way, depending on the length of the journey.

For more local journeys there are the small minibuses, which are called dolmuş (not to be confused with dolma which are stuffed vine leaves!), and these can usually be picked up at any point along their route by flagging them down, but to be on the safe side, try an official stop, designated by a D sign, or wait next to locals who look like they are waiting for a bus. Again, the final destination is usually displayed at the front of the dolmuş. A more unusual system exists in some cities, whereby you must buy your ticket before getting on the bus from an anonymous looking ticket seller. He's usually wearing a hat and hanging around the main bus stops.

Railways in Turkey are cheap but slow. As they serve very few places they are not recommended.

Taxis

Taxis are reasonably cheap and hiring them for a day can work out cheaper than hiring your own car, except in big cities. You can ask a driver to take you to a site and pick you up later without too much worry.

Hitching

If you are using public transport, it may occasionally be necessary to hitch to get to the more out of the way places, or to avoid a 10km walk, but this is usually no problem, just wave down passing vehicles, and don't stick out your thumb (it's an insult!).

STAYING IN TURKEY

Accommodation

Hotels

Hotels can be found in virtually every town, and vary considerably in standard. Almost all have clean beds, but always check out the toilets. Showers are available in many hotels, but most are of poor quality and hot water is not guaranteed, except in the larger establishments. Generally, good hotels can be found easily in the west, and less frequently in the east. One thing that most hotels have in common though, is that they are relatively cheap. The better hotels cost between £5 and £30 a night rising to international rates in İstanbul. Small hotels and pensions (guest houses) range from £1 to £5 a night.

Camping

Campsites are few and far between, but can be found along the coastline, on the southern shore of Lake Van and in the Turkish Lake District. They usually offer few facilities, although most are allied to restaurants. Apart from this, camping is usually no problem anywhere, but it is sensible to ask permission. One thing that must be borne in mind is that there are packs of feral dogs in some of the more remote areas, which have been known to kill. There are also wolves, bears and leopards, but these generally stay well clear of people! Also, do not camp out in the east of the country (from Birecik eastwards), since there are local bandits.

If you are stuck for accommodation, ask at a local restaurant or tea house. They will usually find somewhere for you to stay, and you are often invited to stay in people's homes when in more remote areas.

Food

Turkish food is highly variable, depending on the establishment. Restaurants usually offer the best choice, with lamb or goat forming the main items, generally in the form of kebabs (Kebap). These kebabs come in all shapes and forms, and are freshly cooked. There will also be an alluring range of side dishes, usually of aubergine, peppers and yoghurt, many of which are delicious and can make up a meal on their own. Soups (Çorba) are also common, with lentil (Meçlemek) being especially tasty, and pastry rolls filled with cheese (Börek) are worth trying. Bread (Ekmek) is given in unlimited quantities, and rice (Pilav) is a common addition. Fish (Balık) can be found along the coasts and lakesides. There are not many puddings available, but rice pudding (Sütlaç) and sweet cakes called Baklava are delicious.

Kebab houses are very common, and generally offer a smaller range of dishes, usually just a couple of types of kebab, a soup and the odd side dish. Street sellers offer a variety of roast nuts, seeds and corn-on-the-cob.

Ice cream and pastry houses (Pastanesi) have a wide range of cakes and biscuits, as well as soft drinks and delicious ice cream (Dondurma), and are a welcome sanctuary in the heat of the day. The top ice cream town is said to be Karamanmaraş.

Tea houses (Çay Evi) can be found everywhere, and offer Turkish tea (Çay) served in small glasses (if you want a larger glass, try asking for a Büyük Çay or Düble Çay), soft drinks, Turkish coffee (Kahve) and a yoghurt drink called Ayran, which is said to be good for an upset stomach.

Beer (Bira) can be found in some restaurants, and there are the odd beer houses (Bira Fıçı) in some towns. The beer is state controlled and only available as a few types of lager, the most common being Tuborg and Efes Pilsen. Fruity wines (Şarap) are available occasionally, but by far the most popular drink is the national drink, Rakı, an aniseed flavoured spirit that is usually mixed with water; drink with care.

Shops do not offer a great variety of food, usually just biscuits (Bisküvi), nuts (Fıstık), bread, a few spreads and occasionally cheese (Peynir). Grocers have a wide range of fruit and vegetables. Check out the water melons in Diyarbakır, they're enormous.

Banks

These can be found in most towns and have opening hours from 08.00 to 17.00 on weekdays, (those in remote areas sometimes close for lunch). However, some banks will not deal with foreign currency until 10.30 because they are waiting for the new daily rate of exchange. Nowadays, most banks will change money and traveller's cheques, and some accept Visa and Eurocheques. Türkiye İş Bankası is usually the best bank and will change all the above, whilst the Garantı Bankası often charges no commission. This often seems to depend on how the person dealing with you feels. Don't forget to take a passport when changing money, and in the smaller towns, expect a cup of tea and a chat with the bank manager.

The international airports have exchange facilities which have the same rates as the banks, and at weekends, the larger hotels may change money at a slightly lower rate.

It is always best to make sure you have sufficient funds for the weekends, since there is no guarantee of being able to exchange money, but it is wise to carry some German marks (or US dollars), since you will find that these will be accepted or changed by all manner of people.

Credit cards are not widely accepted, except in tourist shops and some large hotels, with Visa being the most popular.

The People

Turks are among the friendliest people in Europe, and contrary to the cinematic portrayal of the 'shifty Turk', most will go to great lengths to help you. You may get some hassle in towns and cities from carpet salesmen, but they are usually no problem.

Virtually all the Turks you will meet will be men, because although the country is officially a secular state, over 95% of the population are Muslim, and the women accept a subordinate role, staying in the home. Having said this, Turkey is unrivalled in Europe in its equal opportunities programme. There is a high unemployment rate and many men spend their days hanging around in tea gardens.

The Turks have one abiding passion, and that is football. It is shown continually on Turkish TV (or so it seems!), along with a host of other sports, and you will make a lot of friends if you keep a note of the fate of their national side and league teams in their European matches. They also watch matches from all over Europe and are familiar with many British teams. Most Turks smoke, and this can be annoying for the non-smoker. Cigarettes will be offered everywhere. To refuse, just tap on your chest a couple of times with the palm of the hand and don't worry about offending by refusing, since most non-smokers are regarded in high esteem.

Women should be careful about travelling alone, since the Turkish men can be overzealous in their attentions, but if you dress sensibly and act firmly towards them, they will usually go away. Men should not wear shorts away from main tourist areas, as this could offend.

In the west of the country there are usually no problems with birdwatching as the people are familiar with tourists, but in the east, and especially the southeast, some precautions are necessary. The eastern part of the country, along with northern Syria, Iraq and western Iran are part of Kurdistan, and the Kurdish people are not fully recognised by the Turkish administration. This has caused civil unrest in the area and a few freedom fighters operate, mainly out of Syria and Iraq. Their targets are usually military, and the problem is mainly restricted to the area south and east of Batman, so this does not affect many birding sites, but the area around Cizre is not easy to work (see Site Section for more details). Also, there are bandits who rob and occasionally kill both Turks and tourists in the remote mountain areas in the east, so care must be taken. In 1994 most of the east, including the Van basin, Doğubeyazıt, and anywhere east of Diyarbakır, was off limits. Anyone from the UK intending to visit this region should get advice from the Foreign Office (0171 270 4129).

Soldiers can be found anywhere, and are your friends (as long as you don't bird around military zones), since most are young conscripts from the west, so if you have problems contact them or the police. Police are usually friendly and helpful, and cigarettes (especially Marlboro) will help to get them on your side. A cautionary note: don't have anything to do with drugs in Turkey, as the penalties are severe, as anyone who has seen Midnight Express will testify.

Language

Turkish is the official language, and is these days written in Roman script, rather than the old Arabic. Kurdish is spoken between the Kurds, but most speak Turkish as well. A few words of Turkish will go down well in any situation and may get you lower prices at hotels, restaurants etc. A lot of Turks work abroad in the winter, mostly in Germany, and so German is quite commonly understood, with at least one person in each village usually having some knowledge. English is understood in tourist towns by most tourist shops and large hotels. French is rarely understood. In the town of Kulu, a large percentage of the population work in Sweden, and consequently, the most common second language in the town is Swedish!

Turkish is not a simple language at first, but it is very logical and easy to get to grips with after the initial problems. One of the more difficult problems is the pronunciation, and Turks may not understand if you do not speak correctly since there are similar words (eg Kuş (bird) and Kış (winter)) which are pronounced only subtly differently. All letters always have the same sound, with consonants being largely the same as in English, the following guide may be useful.

A	As in Cap	C	As J in Jump
Ç	Like ch in Chat	E	As in Get
G	As in Get	ğ	Lengthens preceding vowel, (eg Ereğli = Ereyli)
İ	(dotted i) As in Pin	I	(undotted i) As e in Wanted
J	As s in Pleasure	O	As in Poet
ö	As u in Murder	Ş	As sh in Shut
U	As in Put	ü	As oo in Boot

Below are some useful words and phrases:

Yes	Evet
No	Yok, Hayır
Hello	Merhaba
Goodbye	Allahaısmarladık (when leaving) Güle Güle (when staying)
Good Morning/afternoon	Günaydın
Good Evening	İyi akşamlar
Please	Lütfen
Thank you	Teşekkür (ederim)
Hotel	Otel
Bus Station	Otogar
Railway Station	İstasyon
Airport	Havaalânı
Marsh	Bataklık, Sazlık
Lake	Göl
Forest	Orman
Sea	Deniz
Bird	Kuş
I am looking at birds/ducks	Ben kuşları/ördekleri bakıyorum

CLIMATE AND CLOTHING

The clothing required is variable with the time of year and the habitat. Good walking shoes are essential, along with light clothing (preferably cotton) for the summer, and a jumper for the evenings in the east, where it can get quite chilly because of the altitude. If you are going to any alpine areas, warm clothing and a waterproof are essential, especially on the Black Sea Coast, where rain is fairly frequent. Sunglasses and a hat are recommended, along with a long sleeved shirt to prevent sunburn and sunstroke in the hotter areas. Remember, that sunburn is easier to get at higher altitudes, so although it is cooler at Van than Diyarbakır, you will burn more readily. In winter, warm clothing is needed as it gets very cold in some areas.

Penduline Tit

HEALTH AND MEDICAL FACILITIES

The chance of catching any serious disease in Turkey is fairly small if adequate precautions are taken. No inoculations are required by law, but it is recommended that you are immunized against typhoid, polio, tetanus and hepatitis A, and some doctors also recommend cholera as well. Malaria is rare, but does occur in southern localities in summer, so a course of malaria tablets is a necessary precaution.

Turkish Tummy

As usual, the most common complaint is Turkish Tummy. It varies in severity from simple diarrhoea to gastroenteritis and is generally caused by the change in diet. Some precautions can be taken to avoid it, such as only eating well-cooked meat, washed and peeled fruit and vegetables and only drinking bottled water, but this does not always work. The best cure is to rest for a day or two and drink plenty of water (containing rehydration salts such as Diarolyte). The symptoms can be alleviated by tablets such as Diacalm. If symptoms are very bad or persistent, consult a doctor or go to a hospital.

Sunburn

Sunburn and sunstroke should be avoided. Take a long-sleeved shirt, hat, sunglasses and use an effective high factor suncream, especially for the first few days.

Bites

Mosquito and other insect bites can be a nuisance, so it is a good idea to take a repellant, and some antihistamine cream. The only other problems you might encounter are bites and even licks from animals, especially dogs, which should be treated by a doctor immediately due to the rabies risk. The same applies to snake bites and scorpion stings (both are very unlikely).

Doctors and pharmacies

Doctors (Doktor) can be found in nearly all towns, and many also have good hospitals (Hastane). Pharmacies (Eczane) are also common and stock a wide range of drugs. Make sure you have adequate medical insurance to cover all medical costs.

MAPS

There are several available maps, many of which cover only the western part of Turkey, but the following are the best:
Roger Lascelles: Turkey, East: 1: 800 000
Roger Lascelles: Turkey, West: 1: 800 000

Also, there are Tactical Pilotage Maps available, by the US Defense Mapping Agency, which are of 1:500 000 scale, and excellent for relief and finding inconspicuous tracks. The problem with them is that the road classes are not shown, and because they show airports and military airfields, you could be accused of spying if found with them. They are therefore best for home use. The Lonely Planet guides to Turkey and Trekking in Turkey are packed full of accurate and well-written information.

All the above are available, along with several other travel guides at Stanfords International Map Centre, 12-14 Long Acre, Covent Garden, London WC2E 9LP (Tel: 0171 836 1321).

White-headed Duck

WHEN TO GO

Birdwatching can be rewarding all year round and which species you want to see is likely to determine the time to visit. The most productive period is from April to October, with both ends of this period being superb for migration. For spectacular raptor passage, the Bosphorus, Borçka and Belen are excellent from late August to early October, while most of the specialities are still present up to the end of September. A less obvious migration takes place from April to May. In Autumn, expect to see Black Kite, Honey Buzzard, Steppe Buzzard, Steppe Eagle (in the East), Spotted Eagle, Lesser Spotted Eagle and Levant Sparrowhawk along with smaller numbers of other species depending on location.

The best period for resident and summering species is the end of April to July, when most species are singing and therefore easier to locate. Bear in mind that species such as Cretzschmar's Bunting, Black-headed Bunting, Blue-cheeked Bee-eater and Pale Rock Sparrow migrate from late July onwards and are very difficult to find thereafter, and Yellow-throated Sparrow is hard to locate when not singing due to its low numbers.

Waders pass through from April to September, with Little Stint being the predominant species, and smaller numbers of Temminck's Stint, Green, Wood and Common Sandpipers, Marsh Sandpiper, Spotted Redshank. The elusive Great Snipe, Terek Sandpiper and Broad-billed Sandpiper can also be found with luck.

Little is known about wintering species in some areas. The east and mountainous areas are under snow and have sub-zero temperatures throughout winter, so much useful information could be gathered by anyone willing to brave the elements. Good numbers of wildfowl can be seen in wetland areas and there is a spectacular gathering of most of the world's population of White-headed Duck at Burdur Gölü from November to March.

Temperatures are warm to hot (20-35°C) from May to September, although it can be decidedly chilly in the east at either end of this period. The area around the Syrian border gets very hot during the day from May to August, and this usually restricts birding to the morning and evening, which are the best times for locating birds anyway. Elsewhere, it is humid on the Mediterranean coast throughout summer, and quite hot on the central plateau. At no time of year are temperatures ideal throughout the country.

Any time of year can be interesting in Turkey, with April to September having the greatest variety of species. Winter is only for those of an inquisitive nature and sound constitution. Check out the Site Information for the best times of year at each site.

INTRODUCTION TO THE SITE INFORMATION

The country has been split into seven regions, each of which have similar species throughout the area, and relate to the zoogeographic regions as outlined in Sandgrouse (1986). There are specific sites in all regions, but the same species could easily be found outside of these sites in the correct habitat, as much of the country has not been studied. A resume of the habitat type for each region is included in its introduction.

The species mentioned are not the only birds present, but are those of most interest to European birdwatchers. A large number of migratory species could also occur at any site, since Turkey is one of the main migratory flyways between Northern Europe and Africa.

The maps are mainly based on the personal experience of the authors, and should therefore be fairly accurate, although some smaller tracks are subject to seasonal changes. There may be a small degree of inaccuracy in some distances, but landmarks are usually included to negate this problem.

Thrace

This is the European part of Turkey. It is mainly agricultural, with a few wetland areas along the coasts of the Marmara and Black Seas. Typical birds of the area include Lesser Kestrel, Roller and Bee-eater.

İstanbul

The ancient city of İstanbul straddles the Bosphorus, the border between Europe and Asia. This bottleneck is one of the most concentrated areas of raptor migration from Europe to the Middle East and Africa, and birds can be seen wheeling above the city in autumn.

Location

İstanbul is easily reached by air from virtually any city in Europe and is also accessible by road and rail. Being a major city it has an extensive, if somewhat chaotic, transport system. The major migration watchpoint is Çamlıca Hill on the Asian side of the Bosphorus. To get there from the city, take the old bridge across the Bosphorus, leaving the road at the exit signposted to Çamlıca. Follow the signs for about a kilometre until the base of the hill (at this point, Çamlıca is signposted to the right away from the hill), and take either of the roads on the left to get to the top of the hill. There are buses to Çamlıca, and the taxi ride from the city costs about £4.

Accommodation

Hotels are plentiful in İstanbul, with prices and quality to suit all tastes. The Aksaray area has several of good quality and value, with good access to the city and transport. The tourist information office at the airport gives good advice on finding and reserving accommodation.

Strategy

Being a raptor migration watchpoint, the best times of year are spring and autumn, with September being the best month. Either of the cafés on the hill provide good watchpoints and refreshments. The best time of day is from mid-morning to early afternoon. The forest on the south and east flanks of the hill holds passerines and is worth a look.

Birds

Storks and raptors are the obvious birds here, especially White Storks and Honey Buzzards. Most of the other European raptors can be seen passing through, with Levant Sparrowhawk being a speciality. Middle Spotted Woodpecker and Sardinian Warbler both breed in the area, and are not easy to find elsewhere in Turkey, while migrating warblers and flycatchers can be seen at the appropriate times. The city itself has plentiful Black Kites and Alpine Swifts and also one of the more local species - Olive-tree Warbler. Look for this species in green areas in and around the city, but it can be surprisingly elusive at times.

Central Anatolia

Central Anatolia is a raised plateau of about 1000m and consists mainly of sparsely vegetated steppe and farmland and is surrounded by mountains. The area is dominated by the huge salt lake of Tuz Gölü, and has a number of seasonal wetland complexes. The summers are hot and the winters cold, and the best time to visit is from late April to September. Typical species which occur throughout the area include Quail, Roller, Short-toed, Lesser Short-toed and Calandra Larks, Tawny Pipit , Lesser Grey Shrike, Isabelline Wheatear, and Black-headed and Corn Buntings. Raptors can be seen anywhere, the most common being Lesser Kestrel, Long-legged Buzzard and Booted and Short-toed Eagles.

Soğuksu National Park

The park is set in wooded mountains and contains a wide variety of woodland birds. It is also a migration stopover site for many raptors and passerines, with almost every species of raptor that occurs in Turkey having been recorded here.

Location

The national park is situated just west of the town of Kızılcahamam, which is on the main İstanbul to Ankara highway about 75km north of Ankara. The park itself is a basin surrounded by high mountains (1400-1700m) with wooded slopes. To reach the park, take the exit off the highway marked Kızılcahamam and follow the main road through the town to the entrance of the park (an entrance fee may be charged). Carry on into the park and follow the circular route (marked Çevre Yolu) which is signposted occasionally. The start of this route is before the end of the metalled road, and to get to it, you must turn left off the road by a restaurant on the left (this should take you past a police hut on the right), and the road snakes up into the hills. There are occasional junctions on this road, but only those which leave the park area have signs for the circle route. If using public transport, there are several buses to Kızılcahamam from İstanbul and Ankara, and it is possible to walk to the park from the town. Once in the park, you can either try to hitch up to the top of the basin or walk up. Most species can be seen throughout the park, but it is a good idea to gain some altitude to watch for raptors. The road across the saddle at the head of the main valley is a good watchpoint.

Accommodation

There are cheap hotels in Kızılcahamam, and a first-class hotel, the Çam, within the park boundary. Camping is permitted within the park, and allows you to listen for owls at night. There are restaurants in the town, but the best one is in the Çam Hotel in the park itself.

Strategy

The best time to visit the site is between April and September, with the largest number of migrant raptors being found in September. During winter, the area is generally under snow cover, but as with most other sites it has been little explored during that season. There is a festival in the park during the last week in August, so avoid visiting it then.

The site can be covered in one day either by car or on foot, but is worth at least two. Early mornings are the best time for passerines, with the raptors generally appearing about two hours after sunrise. Therefore, start with the lower areas at dawn and work up towards a viewpoint to look for the raptors. It is best to look for raptors from the southern slopes due to the sun. The higher slopes can then be explored through the day, and the raptors watched again towards dusk as they come in to roost. A slow journey down after dark may produce owls.

Birds

Black Stork is regularly seen throughout the breeding season, either soaring above the park, or in the river east of the highway. Of the raptors, breeding species include Black Kite, Lammergeier, Egyptian, Griffon and Black Vultures, Short-toed, Golden and possibly Imperial Eagles, Goshawk, Sparrowhawk and Buzzard. Lesser Kestrel, Kestrel and Hobby breed, while Peregrine and Lanner probably breed. During migration, Red Kite, White-tailed Eagle, Levant Sparrowhawk, Lesser Spotted, Spotted, Steppe and Bonelli's Eagles, Red-footed Falcon and Saker have all been recorded. A good place to look for the vultures is the slaughter house, which is west off the highway about 1km south of the Kızılcahamam junction.

The owls present in the area include Scops, Eagle, Little, Tawny, Long-eared and Tengmalm's. The latter is one of only four records from Turkey and is worthy of further investigation. Another unusual record for Turkey is that of Grey-headed Woodpecker. Great Spotted and Green Woodpeckers are common on the hillsides, and Syrian and Lesser Spotted Woodpeckers can be found in the lower areas around the town. Wryneck also probably breeds, but is difficult to locate, and Black Woodpecker has been recorded.

Passerines present are mostly Central European species, but one of the site's specialities is Krüper's Nuthatch. They are widespread and best seen before May and after August, when they are in good plumage. Nuthatch and Rock Nuthatch also occur, and Short-toed Treecreeper is more common than Treecreeper. The Redstarts here are of the race *samamisicus* with white wing patches. Chiffchaff is by far the commonest warbler, but Wood, Bonelli's and Olivaceous Warblers can also be seen. Goldcrest, Blackcap, Whitethroat and Lesser Whitethroat breed, while Orphean, Sardinian, Barred, Garden and Olive-tree Warblers are infrequent visitors. Sombre Tit can be found among the more common species, and *Ficedula* flycatchers should be carefully checked during migration, since any of the European species could be found. Rock Sparrow is occasional, and among the commoner finches you may find Hawfinch and Common Rosefinch. Rock and Ortolan Buntings are both present.

Other Wildlife

Bears, Wolves and Wild Boar are found in the area but are difficult to locate, while Fox, Hare, Field Vole, Lesser Molerat and squirrel species are more common.

Kulu Gölü

This is a typical steppe lake which is highly saline and permanent all year round. It is of great importance as a stopover site for several species of waders and wildfowl on migration and has good breeding numbers of ducks, waders, gulls and terns in summer.

Location

Kulu (or Düden) Gölü is a small lake situated just east of Kulu, which lies 108km south of Ankara on the Ankara to Konya road. The lake is approached by turning east off the Ankara to Konya road at a kebab house in Kulu and following this road to a roundabout, which has the Atatürk statue in the centre of it. Take the exit directly across (the second) past the Türkiye İş Bankası on your right to reach the lake. Follow this road out of the town past a wall on your right. The track continues to the southern end of the lake, where there is a small, almost isolated pool, and then away across the hills towards Tuz Gölü. To reach the northern end of the lake, take the third exit at the roundabout which will take you through to an open area. Cross a bridge and then carry straight on, or alternatively, turn right and then left opposite the school and either track will take you to the lake. It is possible to drive round the northern end of the lake and down some of the eastern edge.

Accommodation

There is a hotel in Kulu, but it is not recommended. There are several hotels in Ankara and Konya, with variable prices, which are both just over an hour away from Kulu. There are several cheap restaurants in Kulu, and some more expensive ones along the road from Ankara to Kulu. Camping is usually no problem around the lake, but ask permission first from anyone who looks like they can give it. Do beware of the large Anatolian sheepdogs with their spiked collars.

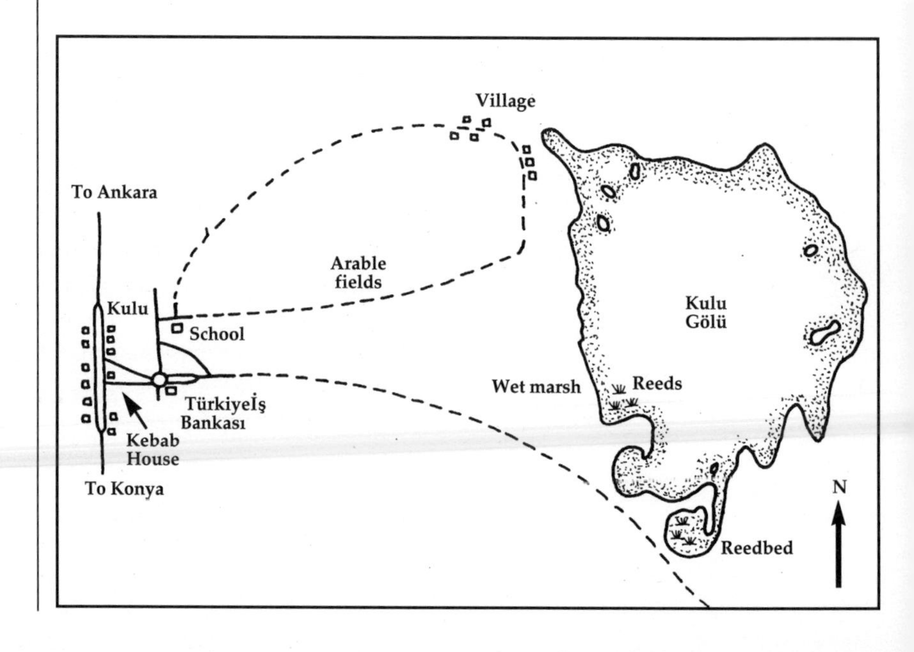

Strategy

The site can be visited at any time of day, and is easily worked in a day. Weekdays are best, since people picnic on the lakeside and make a lot of noise at weekends, although this is not really a major problem. The birds can be located anywhere around the lake, but the only reedbeds are a small patch on the western shore and in the small nearly isolated pool at the southern end of the lake.

It is possible to drive to either end of the lake, and part of the way along the eastern shore, and a car serves as a good hide. You can walk all the way round the lake in a day and this will give the best chance of seeing all the species present, although it is a very long walk.

Birds

The lake holds the largest breeding population of White-headed Duck (c.40 pairs) in the country, and good numbers of Ruddy Shelduck, Shelduck, Gadwall and Red-crested Pochard. On the lakeside marshes Oystercatcher, Black-winged Stilt, Avocet, Lapwing and Redshank breed. Black-necked Grebes are much in evidence, with smaller numbers of Little and Great Crested Grebes.

The reedbeds hold Squacco Heron, and Moustached and Great Reed Warblers. The surrounding fields and steppe have Quail and Black-headed Wagtail and if you are lucky, you may find Great Bustard. Greater Sand Plover can be found around the lake fringes, especially at the northern end, along with larger numbers of Kentish and Little Ringed Plovers. Mediterranean and Slender-billed Gulls breed among the more common Black-headed Gulls, and Gull-billed Tern and Collared Pratincole are usually present.

Other birds present throughout the summer, usually just visiting to feed, include White Pelican, Great White Egret, Spoonbill, Greater Flamingo and Black-bellied Sandgrouse. Marsh and Montagu's Harriers, Lesser Kestrel and the occasional Peregrine hunt around the lake.

Greater Sand Plovers

During spring passage, Teal, Pintail, Shoveler, Pochard, Tufted Duck, Spur-winged Plover, Little and Temminck's Stints, Curlew Sandpiper, Ruff, Marsh, Wood, Green and Common Sandpipers, Greenshank, Spotted Redshank, Black-tailed Godwit, Turnstone, Red-necked Phalarope, Little Gull and Common, Whiskered, Black and White-winged Black Terns all stop over, and occasionally over-summer, while Red-throated Pipits pass through in numbers.

Autumn passage has basically similar species, with huge numbers of Garganey and Little Stint. Rarely recorded species include Demoiselle Crane, Lesser Sand Plover, Slender-billed Curlew, Great Black-headed Gull and large flocks of Rose-coloured Starling. Skuas are also surprising species to find in autumn, with Long-tailed and Arctic both recorded.

Other wildlife

Kulu has some good remnants of Anatolian steppe particularly on the eastern shore. The flowering species of the steppe will only be of interest to the botanist, however it does give a good feel for a habitat which is increasingly rare in the region. An unusual lizard, *Agama ruderata*, can be found here and more commonly at the nearby Tuz Gölü where it frequents the edges of the saltflats. A large jerboa with an extremely long tail is quite common in the area but is normally only seen at night.

Ereğli Marshes

The large area of marshes west of Ereğli offers some of the best birding in Central Anatolia, with astronomical numbers of birds being present from April to July when the lake is at its highest. The main lake (Akgöl) is highly seasonal, and by September, the little water left there (at the southern edge near Böğecik) is packed with birds. Hotamış Marshes, which are just south of the Ereğli to Konya road and 30km west of Ereğli Marshes used to be similarly good for birds, and were the only known breeding locality of White-tailed Plover in the country. The marshes have been largely drained, but the area is still worthy of exploration.

Location

The lake can be approached from two directions. To get to the southern end, which has quite dense reedbeds, take the minor road to Böğecik off the Ereğli to Karaman road about 18km west of Ereğli. Just inside the village, you come to a crossroads. Take the right turn which skirts a hill and you should see the lake on the right of the track after a few kilometres. To get to the northwestern edge of the lake take the road north off the Ereğli to Karaman road just before the 'Ereğli 17' signpost (or just after the 'Ayrancı 22' signpost from the opposite direction) opposite a quarry. This road will bring you to the village of Adabağ after 4km, and Tatlıkuyu after 9km. In Tatlıkuyu, turn left at the post office (PTT) and stork's nest and cross a ditch at the edge of the village. Take the left hand most track after the ditch and follow this until you reach a water pump. Take the left fork past this water pump,

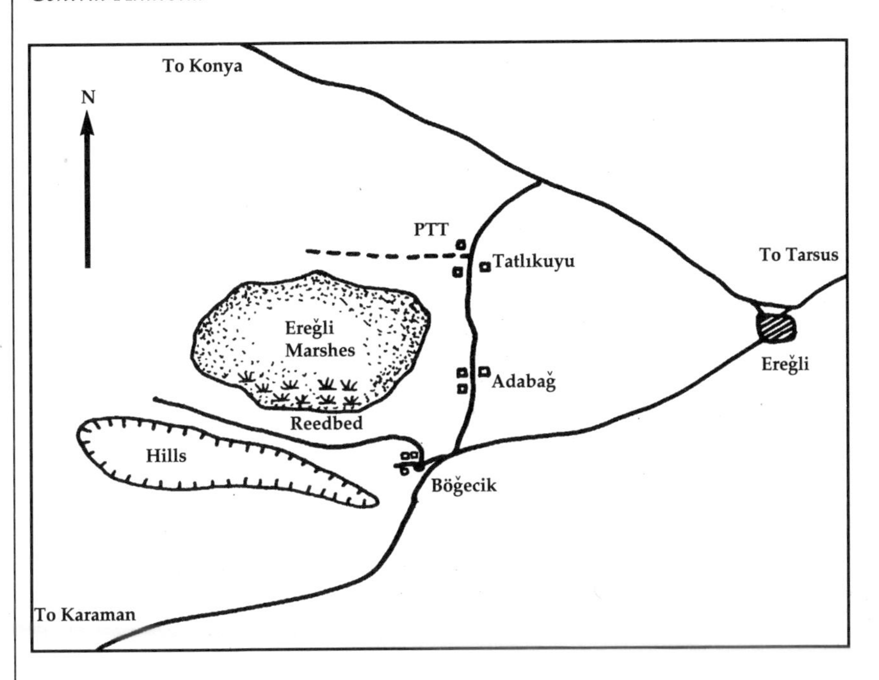

and after a few kilometres you should reach the lake edge. The tracks mentioned are not easy to find, but if you head west-southwest from Tatlıkuyu, you will reach the lake eventually.

Accommodation The nearest hotels are in Ereğli, and are of a reasonable standard. Camping around the lake is probably possible away from the fields, but ask first. There are restaurants in Ereğli and a tea house in Tatlıkuyu.

Strategy Spring and early summer are the best times to visit the site because the water level is at its highest and several areas are visible. In autumn (and sometimes also in spring), the only reachable water is at the southern end of the lake and is mostly obscured by reedbeds.

The northern end of the lake offers the best viewing for flamingos, pelicans, waders and open water species in spring, while the more secretive species will frequent the southern reed areas. In autumn, it is a long trek from the north across soggy ground to get to the lake and it may be necessary to walk through the reedbeds to see birds at the southern end.

It is worth spending a few days at the lake, though most species can be seen in one day with a vehicle. Sudden storms can make the dirt tracks impassable in minutes, so be careful when driving around the lake area.

Birds Wintering species include good numbers of herons, geese and ducks but these are under severe hunting pressure.

Migration starts in April, with White and Dalmatian Pelicans passing through in good numbers (White are much the commoner), passage ducks and waders including Marsh Sandpiper, Little and Temminck's Stints and the occasional Great Snipe and Terek and Broad-billed

Sandpipers. Large numbers of Spoonbills and Great White and Little Egrets can be seen in the open water, with large flocks of Greater Flamingo feeding among them.

During the summer, pelicans are usually present, but Dalmatian can be difficult to see. All the Turkish herons breed with the exception of Grey (which is usually present) and Bittern. Cranes are regularly seen and Pygmy Cormorants are common, as well as the occasional Cormorant. The commonest duck is Red-crested Pochard, although Ruddy Shelduck, Shelduck, and small numbers of White-headed Duck and Marbled Teal also breed. This is also one of the few sites in Turkey for Mute Swan.

Raptors present include Short-toed Eagle, Marsh and Montagu's Harriers and Egyptian Vulture, although none except the harriers are guaranteed. Saker and Eagle Owl once bred in the area and may be found again with luck. There is a colony of rather tame Lesser Kestrels in the school grounds at Adabağ.

Around the fringes of the lake, Black-winged Stilt, Avocet, Collared Pratincole, Little Ringed, Kentish and Spur-winged Plovers can be seen. More elusive are Stone-curlew and Greater Sand Plover (over 100 have been seen in one group). Slender-billed Gulls breed, along with Gull-billed, Little, Whiskered and Black Terns. Black-headed Gull and White-winged Black Tern can also be seen.

Reedbed passerines are not easy to see, but Penduline and Bearded Tits both occur, along with Reed, Great Reed and Moustached Warblers. Black-headed Wagtails are common around the lake along with the usual commoner Anatolian species.

In autumn, the same migratory species as in spring can be seen, but watch for Citrine Wagtails among the flocks of Yellow Wagtails, and look out for Spotted, Baillon's and Little Crakes and Water Rail along the reed fringes.

Other wildlife

Susliks are quite common on the central plateau and here you will have a chance to get to know one Suslik 'town' quite well as it occupies a favourite watchpoint by the track east of Boğecik. Here a small mound is riddled with holes and the Susliks poke out their heads and give a warning whistle. Watch out for the scorpions which also live on the mound.

Sultan Marshes

The Sultan marshes are justly renowned as one of Turkey's premier wetlands. A vast area of marshes, mud and salt flats, reedbeds, canals and both freshwater and salt lakes nestle in a basin below the vast bulk of Erciyes Dağı (3917m) - the Mount Argaeus of the classical era and one of Turkey's highest peaks.

Location

Sultan marshes are situated some 60km south of Kayseri and a similar distance southeast of Urgüp (Cappodocia). The basin is ringed by roads, with the town of Develi at the northeastern corner. The area is

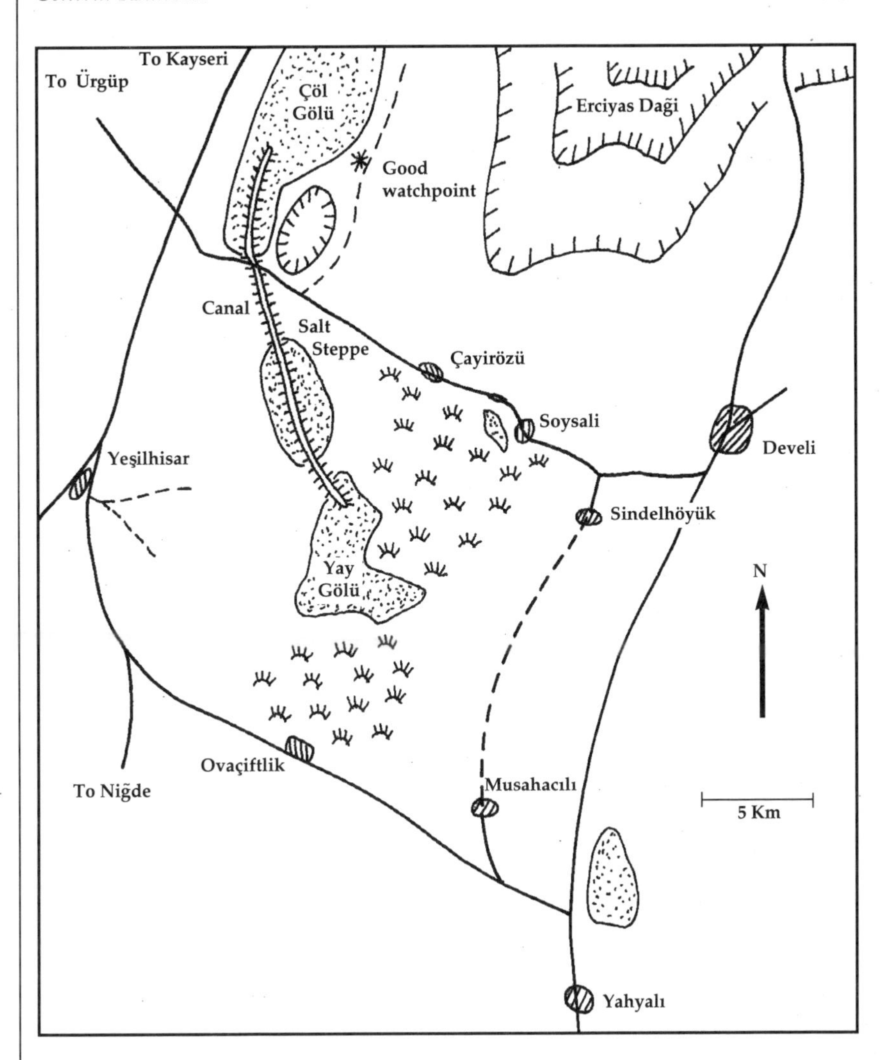

well served by buses and dolmuses, and Kayseri has an airport. There is a high pass between Develi and Kayseri over the shoulder of Erciyas Dağı which passes through Kayak Evi, a small ski centre at 2100m, should one want to try some high altitude birdwatching!. A word of warning, Erciyas Dağı is a dangerous mountain and attempting to reach the highest areas should only be considered by experienced mountaineers. The summit area is eroding at an alarming rate and the terrain is highly unstable. Hasan Dağı, about 100km to the west is much easier and also has more wildlife including Wolves and marmots.

Accommodation

There are plenty of hotels to suit all tastes in Kayseri and Cappadocia. More locally there are a couple of standard hotels in Yahyalı, at the southeastern corner of the marshes. There are restaurants in Yeşilhisar, Develi and Yahyalı.

Strategy

Sultan marshes provide a birding spectacle all year round. April through to June is an excellent time to visit with a wide variety of migrants passing through, and also some spectacular concentrations of breeding species. In winter there is a vast array of birds, sometimes as many as half a million, although species diversity is not as high as in spring and autumn. It can get very cold in winter and large areas of the marshes are often frozen, although the saline Yay Gölü nearly always remains free of ice.

Access is a problem over much of the area because of variable water levels, the low density of roads and drivable tracks in the area, and from the sheer size of the marshes. A good starting point for the southern part of the marshes is at Ovaçiftlik where there is small information stand on the wildlife of the marshes and an observation tower which gives a good overview to the north. There is excellent birdwatching from the road in either direction from Ovaçiftlik with waders, terns and herons being the most obvious species present.

Further southeast along this road, in Ilyasli, there is a road northwards to Musahacılı. From here, there is a track to Yenihayat and thence to Sindelhöyük. Just north of Yenihayat there is a drainage canal and the banks of this afford a good view over the area. It is possible to get close to Yay Gölü by following the banks of the canal, and there are mounds of earth which are left over from the drain construction which offer decent observation points. The muddy lake fringes are excellent for waders and sometimes flocks of up to 10,000 Greater Flamingos can be found. The tracks in this area are usually drivable in summer, but quickly become impassable after rain. It is possible to take a boat into the reedbeds from Ovaçiftlik. This is the best way to look for some of the rarer reedbed and duck species. It is best to go early in the morning as you will need to be the first boat into the area as later boats tend to follow the same routes. There will be a fee for this, the size of which will depend on how much the local boat owners think you will be prepared to pay!

The road across the northern side of the marshes runs very close to the southern tip of Çöl Gölü, the shores of which are very good for wading species. The saltflats and steppe along the first few kilometres from the main road junction are good areas to look for Greater Sand Plovers, as they often breed very close to the road. There is a large drainage canal which crosses the road and goes right into the centre of Çöl Gölü. This is quite productive and walking along the bank in either direction should ensure a good mix of herons, waders and terns. A track runs along the east side of Çöl Gölü, starting a few kilometres east of the lake just passed a small hill. This will bring you quite close to the lake shore. Walking along this part of the shore can be quite productive for waders, and for a good overview of the lake which often holds Greater Flamingos and pelicans. There are also good areas of marsh to the south of the road between Çayirözü and Soysalı. On the far western side of Sultan marshes there are a number of tracks running eastwards from Yeşilhisar into an area of mostly unimproved steppe.

Birds

Over 250 species of bird have been recorded at the Sultan marshes. The marshes hold very important breeding populations of Pygmy

Pygmy Cormorant

Cormorant, Little Bittern, Night and Squacco Herons, Little Egret, Glossy Ibis and Spoonbill. Pelicans of both species are frequently seen, although only White Pelicans breed regularly. The many small reedbed lakes and channels, especially in the south of the marshes, have nationally or internationally important breeding populations of wildfowl. Most notable of these are Ferruginous and White-headed Ducks. Red-necked Grebe, Gadwall and Red-crested Pochard also breed in these reedbeds. Bearded and Penduline Tits are both reasonably common. During migration large numbers of waders and terns pass through. Some waders stay to breed, including Black-winged Stilt, Avocet and Greater Sand Plover. Slender-billed Gull and Gull-billed Tern both breed or have bred in the area. Roller, Hoopoe, Syrian Woodpecker and Lesser Grey Shrike can be found in the steppe area on the western fringes. In winter a huge number of wildfowl can be found, particularly Wigeon, Teal and large flocks of geese. Over the years, a large number of unusual migrants have been recorded at Sultan marshes with only a limited amount of observer coverage. The potential for finding something out of the ordinary at this site must be quite high.

Other Wildlife

Many species of amphibian and reptile are found here, among them Dice Snake, Grass Snake, terrapins, Marsh Frog and tree frogs. The *Artemesia* and *Bromus* steppes contain many endemic steppeland species of plants and small mammals, and the wetter areas have vast displays of orchids (*Orchis palustris* and *O. laxiflora*) in the late spring and early summer.

The Lake District

The Turkish Lake District consists of a series of deep lakes on the fringes of Central Anatolia. All are contained in the Central Anatolian section for convenience, although only Eber, Akşehir and Çavusçu are in the region. The area is used by Turkish tourists, but is remarkably peaceful, and has been seldom visited by birdwatchers during the breeding season, although winter counts of wildfowl on the lakes are made annually. Of particular note is the beautiful Kovada National Park to the south of Eğirdir which protects a large area of unspoilt mountain forests and karst habitats. There are many high mountain ranges in the area. Two of the most productive are the Barlar Dağı, northwest of Eğirdir and the Dedegöl Dağları, on the west shore of Beyşehir Gölü.

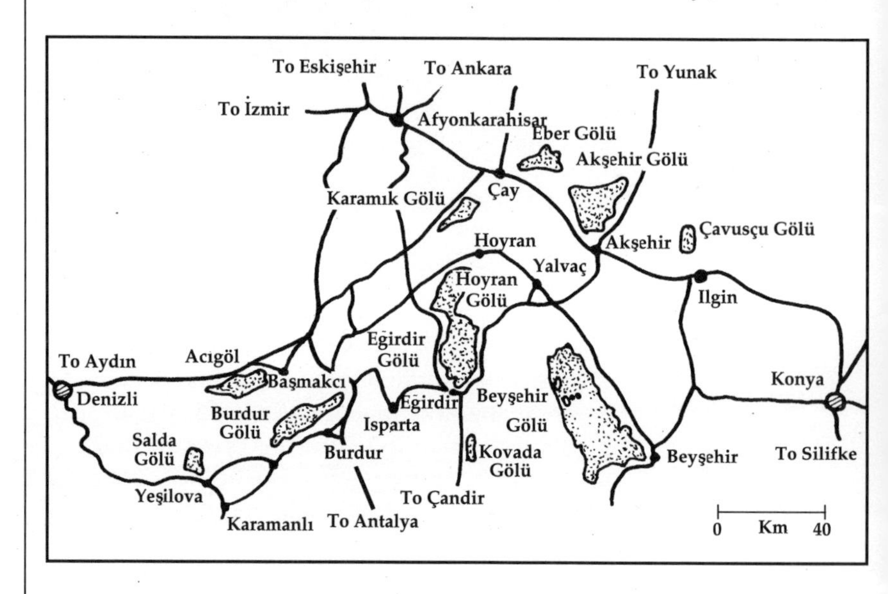

Location

All the lakes have reasonably easy access since they are mostly quite large, but some stretches are difficult to get to except on foot. Akşehir is probably the richest lake in the area for birds but it is also the most difficult to birdwatch due to the wide belt of very tall reeds surrounding much of the lake. The best access is to take the Yunak road from Akşehir town until you reach the village of Orkaköy eleven kilometres to the north. Turn left onto a dirt track signposted 'Titiköy 2', and keep left at a junction 400 metres further on. After about two kilometres there is a derelict barn on a slight rise in the ground to the left. This is a good place to leave a vehicle, especially if the ground is at all damp, otherwise continue on about 500 metres to a small derelict hut from where tracks (walkable, but damp) lead through the reeds to little bays where fishermen keep their boats.

Kovada National Park is approached by following the Çandir road south of Eğirdir. To reach the very highest mountains of the area follow the Aksu road southeast of Eğirdir until you reach a junction signposted to Yakaköy some ten kilometres past Aksu. Follow this road to the village, turning right at the official looking building on your right and

Roads
Driveable track

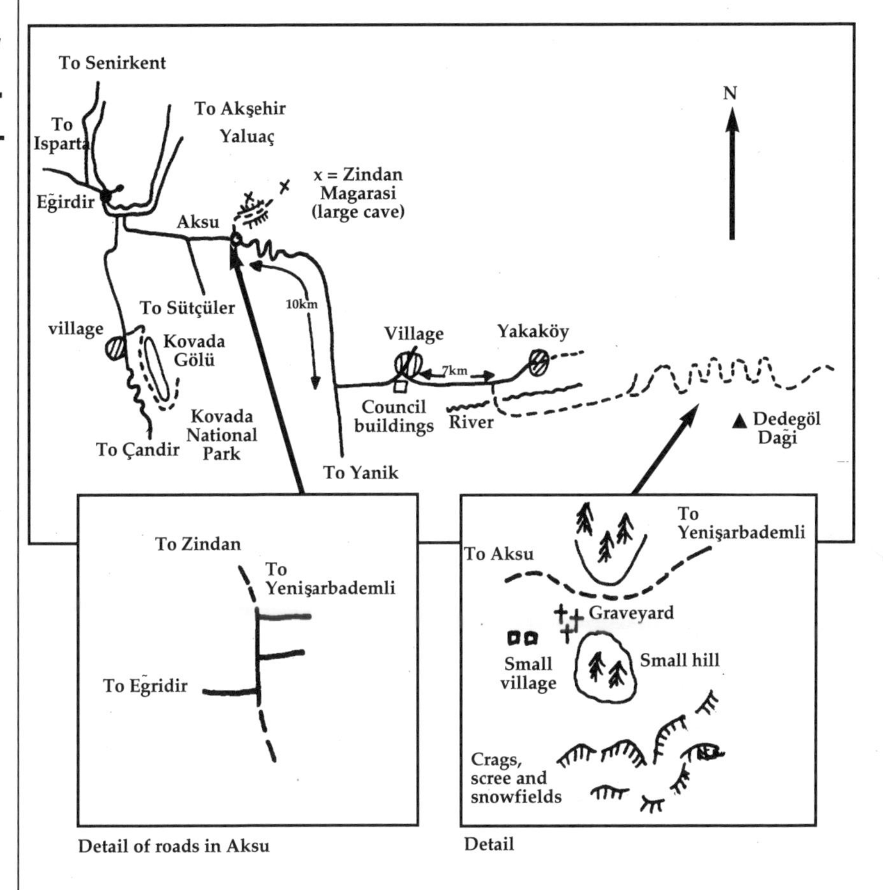

Detail of roads in Aksu

Detail

Key

Roads
Driveable track
Wet tractor tracks

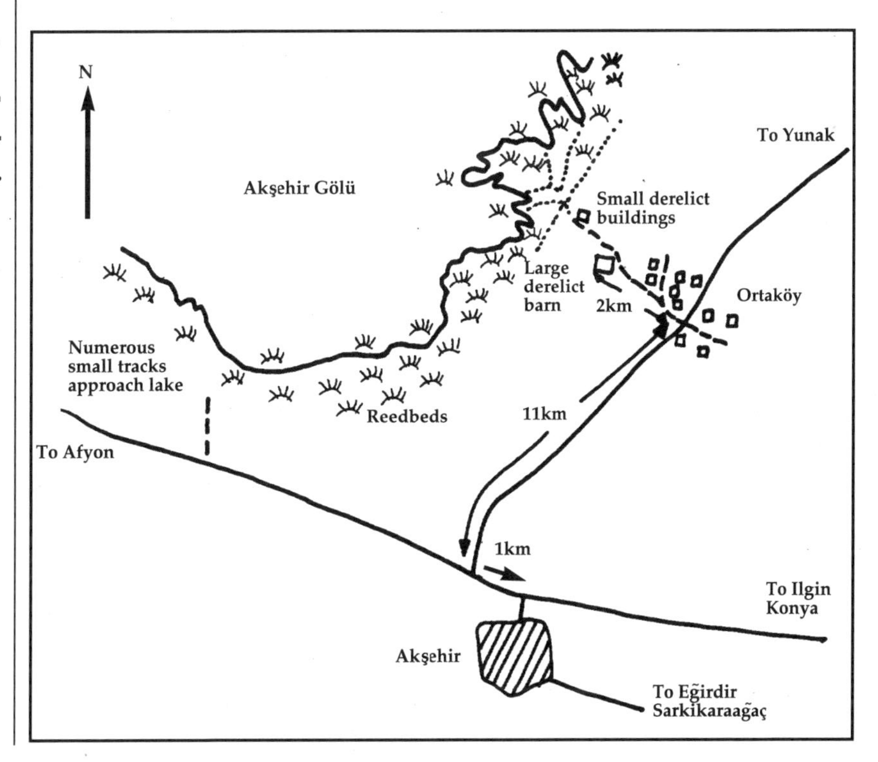

about seven kilometres further on there is a wide flat valley bottom. Here there is a track forking to the right (south) which immediately crosses a bridge, before running east parallel to the river. This is a forest track and not for the fainthearted, but eventually you will top a pass with stunning views down to Beyşehir Gölü below you. The pass is at about 1750 metres and the bare snowy slopes of Dedegöl Dağı are very close by to the south. The track continues on down to Yenışarbademli.

Burdur Gölü is best birdwatched from the southeastern end, where a series of rough tracks run downhill from the main Burdur - Keçiborlu road crossing a railway line near the mudflats which cover a wide area at this end of the lake.

Accommodation

Hotels and restaurants can be found in all the towns in the area, including Konya, Akşehir, Afyon, Burdur and İsparta, and there are campsites on the southern shores of both Eğirdir Gölü and Beyşehir Gölü. Eğirdir is a good base to use for the whole region. There are some very nice pensions on the island offshore of the town and the large lakeside Eğirdir hotel is very good value.

Strategy

Birdwatching in this area is difficult but not impossible without a car. Buses serve all the towns, but it may be necessary to hitch to some of the more out of the way areas. Birdwatching is good all year round, but it is not always easy to get around in winter due to snow. Do bear in mind where the sun is going to be when you visit each site, especially the lakes. Watching from the eastern shore of Akşehir as described above is only practical in the morning as the sun will be in your eyes in the evening, and heat haze can make things difficult at Akşehir, Burdur and Çavuşçu by midday. It is quite a large area and worth a good week to explore, and is therefore ideal for someone based in the district or on holiday in the Antalya region.

Birds

The lakes themselves do not have a very large number of birds on them during the spring and summer, but where there are reedbeds, expect to find any of the heron species, Pygmy Cormorant, Bearded Tit, Reed, Great Reed and Moustached Warbler and Reed Bunting. In trees around the lakes, there are Olivaceous and Cetti's Warblers as well as the commoner Red-backed Shrikes, and the lake fringes hold Tawny Pipit, Lesser Short-toed Lark and Black-headed Wagtail. Akşehir Gölü is good for terns and waders during migration periods, while herons, egrets, Marsh Harriers and Collared Pratincoles are present throughout the summer. Çavuşçu Gölü has started to attract large flocks of White Pelicans in recent years and Whiskered Terns breed in a marsh on the west shore. Burdur Gölü does not attract a wide range of species during the summer months, but there are usually good numbers of Greater Flamingos there and the surrounding fields are good for Roller and Bee-eater.

The surrounding wooded hillsides have Rock and Blue Rock Thrushes, Black-eared and Finsch's Wheatears, Sombre Tit, Krüper's

Nuthatch, Bonelli's, Orphean and Rüppell's Warblers and Cretzschmar's Bunting. This region has a sizeable population of White-throated Robin, and on some of the hills around Eğirdir they are the commonest species. The lower level open scrub and trees hold Syrian Woodpeckers, while higher areas of denser forest will have Great Spotted and if you are very lucky, White-backed Woodpecker. Middle Spotted Woodpecker could be found in any of these areas. Olive-tree Warbler should also occur, but it is rarely seen and worth looking out for.

The highest mountain areas hold good populations of Chough, Alpine Chough, Water Pipit, Black Redstart, Woodlark, Shorelark and Red-fronted Serin. Golden Eagle, Griffon Vulture and Lammergeier can all be seen and Crimson-winged Finch, while probably not common, is reasonably easy to find on Dedegöl Dağı.

Spring migration sees massive movements of Turtle Doves, Ortolan Buntings, Red-backed Shrikes and Bee-eaters among others in a broad front through the area. Also notable are large numbers of Red-footed Falcons passing through in large groups. A good area to look for these is on the wet fields between the village of Ördekçi and Sarki Karaağaç to the north of Beyşehir Gölü. When the weather holds the migration up they can be found in their hundreds here, often in the company of Imperial Eagles.

In winter, the lakes become a haven for wildfowl, and unfortunately a magnet for hunters. Burdur Gölü is the major wintering ground for White-headed Duck in the world with around 11,000 wintering there in some years (over 70% of the world population), along with huge numbers of Black-necked Grebe, White-fronted Goose, Pochard, Red-crested Pochard, Tufted Duck and more Coot than you thought existed. White-tailed Eagle has also been known to winter at Burdur. Expect the same species, but smaller numbers at the other lakes along with several other duck species. Also, many wintering raptors gather around the lakes, with Golden and Imperial Eagles, Black and Griffon Vultures and Lammergeier all having been recorded.

Other Wildlife

The lake district region has a very rich flora, particularly among the bulbous species, and the alpine zone of Barlar Dağı and Dedegöl Daglari are two of the most prolific areas for the botanist. Orchids are well represented with the spectacular Komper's Orchid being quite common to the west of Eğirdir. Watch out for a stunning red tulip, *Tulipa armena*, on the pass over the Sultandağları between Bağkonak and Akşehir. Also of note are a variety of *Fritillarias* which occur in the area, including the endemic *Fritillaria whittalli* which can be seen on Dedegöl Dağı. Anatolian Leopard has been recorded from Kovada National Park, although there are no confirmed recent sightings. Wolf and Bear also occur, but are now very rare. Wild Boar are very common in the park, but actually getting to see them is difficult. You can't miss signs of the their rooting in the woodlands that line the tracks and roads within the park. The lakes on the edge of the central plateau such as Akşehir and Çavuşçu have good populations of Susliks - watch out for one of their 'towns' by the derelict barn near Ortaköy on the shores of Akşehir.

Southeastern Turkey

This region borders the Syrian desert and is a mixture of low lying hills to the west and high mountains to the east. There is a large diversity of birds in the region, but no significant wetlands. The Euphrates and Tigris rivers cut through the area, providing valuable water to the dry environment. Typical birds which are widespread include Roller, Bee-eater, Crested, Short-toed, Bimaculated and Calandra Larks, Tawny Pipit, and Isabelline, Black-eared and Finsch's Wheatears. The commonest raptors are Long-legged Buzzard and Short-toed Eagle.

Birecik & Halfeti

The area around these towns on the banks of the Euphrates is one of the best birding sites in Turkey, with several difficult species being easily found. Their location on the river next to arid hillsides provides a diverse range of habitats, and so many species can be located in just a small area. For example, one of the Western Palaearctic's most elusive birds, (Striated) Bruce's Scops Owl, can be found in the most unlikely location - a tea garden.

Location

Birecik is situated on the main Gaziantep to Urfa road, where it crosses the Euphrates (Firat Nehri) and is well signposted. Halfeti is 35km north of Birecik and is most easily reached by taking a metalled road signposted to Halfeti about 6km east of the Birecik junction. Alternatively, the track which heads north out of Birecik also meets this road.

Birecik is easily reached by coach from Gaziantep or Urfa, and buses to Halfeti run from Birecik.

Birding locations in the area are on the map and explained in the 'Birds' section.

Accommodation

There are a few hotels in Birecik, but none in Halfeti. The Merkâlem Motel is located on the western bank of the river, next to the garages, and is good value with a 24-hour cafeteria. If full, try the Hotel Doğan, which is clean and cheap and situated at the northern edge of town. The Hotel Özsu is located next door to the Doğan.

There are a few restaurants in Birecik, one not far from the Hotel Doğan is good value. You can eat outside in this restaurant, which affords good views over the river - and Pied Kingfishers while you eat! There is an excellent restaurant between the Merkâlem Motel and the river. It is relatively expensive but also has entertainment and a swimming pool.

Strategy

There are a number of places to visit in the area, and it is well worth several days, although all the species can be seen in one or two days, since their locations are so well known. Yeşilce can also be visited from Birecik, as it is only just over one hour drive away.

April to September is the best period to visit the site, but at either end

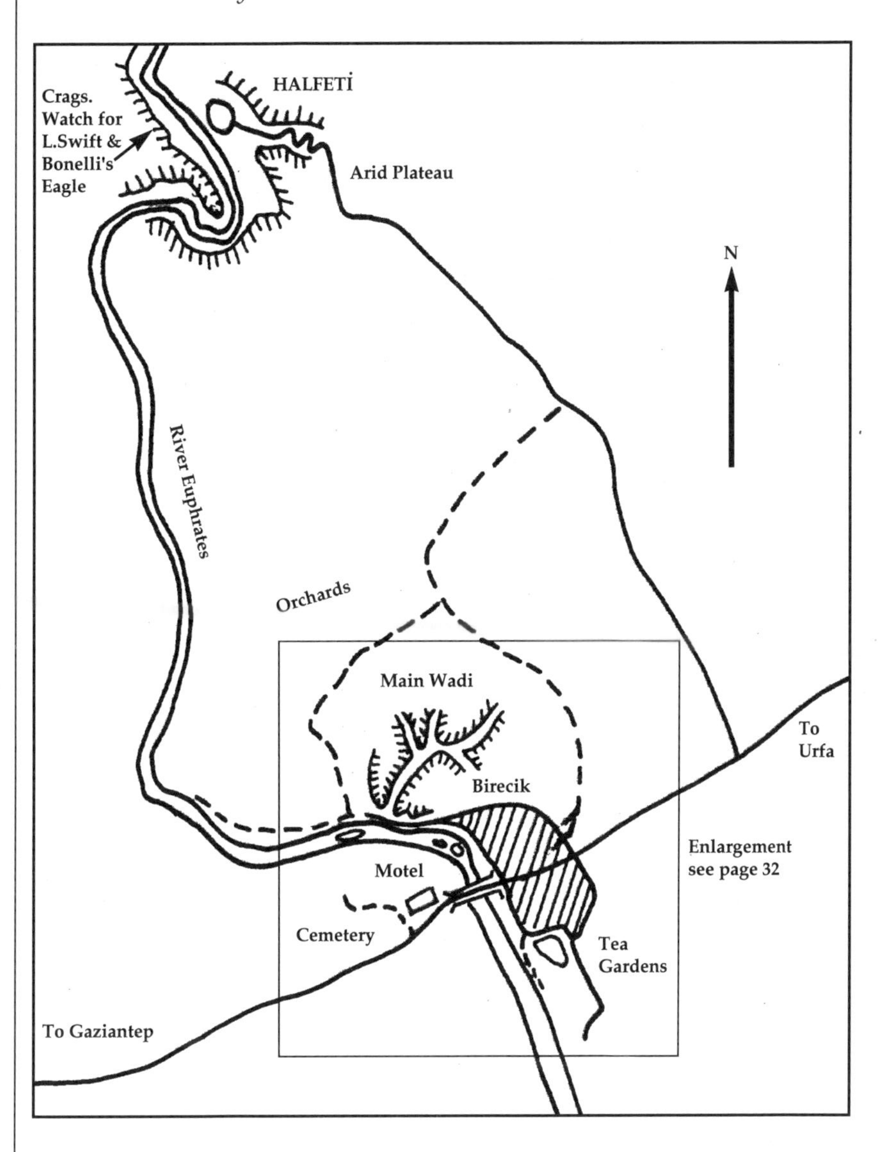

of these times, some of the migratory species such as Blue-cheeked Bee-eater and Yellow-throated Sparrow may not be present. When visiting the area remember that it gets very hot during the middle of the day, which is not good for people or for birds. Birding is best crammed into the first and last few hours of the day. If you are in a hurry, the best places to bird at midday are the orchards north of Birecik and Halfeti, where there is shelter from the sun. Not much is known about the birds present during the winter.

Birds

This section is divided into locations to within the area.

Along the river itself, Pied Kingfishers are regular, and the scrub along the river banks holds Ménétries' and Graceful Warblers. The Graceful Warblers are more common south of the bridge. Palm Doves are common about the town.

Key

New 'Farmers' Market 1

Hotel Doğan 2

Old Arab Market 3

Petrol station 4

Sandpit 5

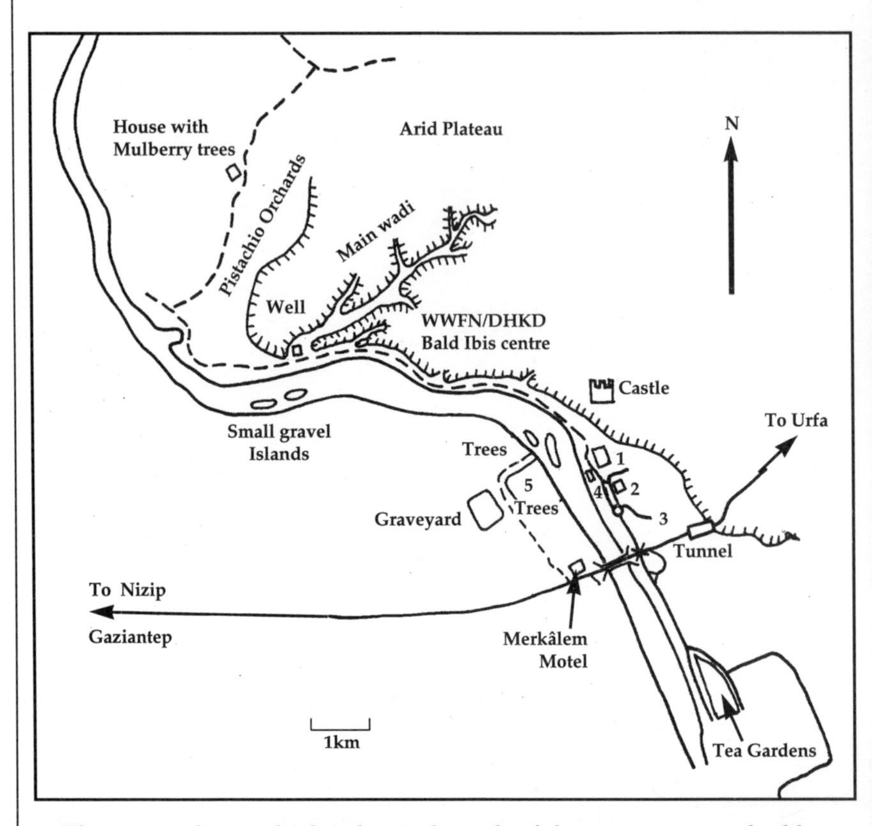

The tea gardens, which is located south of the town, are reached by taking the main road south through town and turning right off this onto a stretch of dual carriageway alongside the river. A few hundred yards along this stretch is a wooded tea garden, which you can park inside. During the day, purchase a drink from the waiter and then inquire about the Striated Scops Owl (Baykuş). He may be able to locate a roosting bird for you (you'll be amazed when he does, they're difficult to find even when pointed out). If you're out of luck, return at dusk and the birds can usually be seen either flying around the tables and trees or in the quieter trees just south of the fenced area. Remember that there are also Scops Owls present, so beware, and Long-eared Owl has also been seen here.

Just west of the river, a track leads up the river from between the garages. About 500m along this track, there is a fenced wooded area on the right. There are sandbanks just inside the fence which have nesting Bee-eaters and Blue-cheeked Bee-eaters, which are easily visible from the road. This colony was badly damaged by sand extraction in 1992. The woods have also been known to hold Long-eared Owl.

There are no longer any wild birds in the Bald Ibis colony but birds can still be seen north of the town. Follow the signposts to the colony (they have ibis pictures on them) north out of town. On this road, just after the new 'farmers' market, the cliffs on the right hold an irregular small colony of Little Swifts. Carry on for about 2km and you will reach the fenced off wadi with the ibis colony on the right. The warden will be happy to show you around.

Striated Scops Owl

Next to the ibis colony, the road crosses a small bridge and there is a well and wadi on the right. This wadi and the surrounding hillsides hold See-see Partridge, Chukar, Desert Lark and Rock Sparrow. There is also an Eagle Owl which roosts a couple of kilometres up the wadi. Cream-coloured Courser are recorded rarely, but with luck, you may find Desert Finch and Pale Rock Sparrow. Opposite this wadi, there are gravelly spits in the river where good-sized flocks of Black-bellied and Pin-tailed Sandgrouse fly in to drink about two hours after dawn.

Another kilometre north of the wadi, the road crosses a small stream and there are pistachio orchards on the right. These orchards have a host of good species including Syrian Woodpecker, Rufous Bush Robin, Olivaceous, Upcher's and Ménétries' Warblers, Spanish, Dead Sea and Yellow-throated Sparrows and Desert Finch. Most of these are easy to find, but Yellow-throated Sparrow is only present in small numbers and is best located in May by song. The best area is around an obvious stand of tall poplars. In these orchards, try to ask permission, keep to the tracks and don't trample crops to avoid annoying the residents.

The arid hillsides just south of Halfeti hold several species, most notably, Rock and Eastern Rock Nuthatches, Cinereous Bunting and the occasional Pale Rock Sparrow and Red-tailed Wheatear.

Just through the village of Halfeti, there are rocky outcrops on the right and orchards on the left. The orchards hold Eastern Rock Nuthatch and Desert Finch, while Striated Scops Owl has also been recorded. On the opposite side of the river, the steep cliffs are home to a small colony of Little Swifts and Bonelli's Eagle nest nearby. Again, when exploring the orchards, ask permission and keep to paths.

Other wildlife

The wadis and semi-desert on the east side of the river support an interesting flora in early spring (March to early April) including several orchids and tulips. The Euphrates Soft-shelled Turtle *Trionyx euphraticus* inhabits the murky waters but is one of several species which are threatened by changes in the water regime caused by the construction of the immense Atatürk barrage scheme further upstream. Scorpions are frequent in the wadis and there are plenty of small fish and frogs in the small pools in the wadi bed. The surrounding hills are a good place to see Jackals in the late evening or early morning, and a night drive up to Halfeti is a good way to see Jerboas kangaroo-hopping across the road. Ceylanpınar, some two hundred kilometres to the east, on the Syrian border, is one of the last places in Turkey where Varan and Goitred Gazelles can be found. This is a sensitive area and access is difficult.

Yeşilce

The Kartal Dağı between Gaziantep and Bahçe are arid hills which contain a wealth of good birds. Small streams keep the valleys lush and allow a wide variety of species to flourish. The valleys at İşikli and Durnalık, near the village of Yeşilce hold all these species and are easily accessible.

Location

The village of Yeşilce is about 20km west of Gaziantep on the main Gaziantep to Bahçe road (Note: A new road is planned which will start east of Gaziantep and bypass the area, so beware. See Lascelles Map of Turkey, East for details). Just west of the village are two tracks heading south, one signposted to İşikli and one to Durnalık (the signs are a bit rusty). Along the İşikli track, drive to the village and park, then search the surrounding hillsides on foot. Along the Durnalık track, follow the road until it bends left and there is a valley heading off to the right. Walk up this valley and search the hillsides and fields until you reach a barren rocky area at the top of the valley.

Buses run regularly along the main road between Yeşilce and Gaziantep.

Accommodation

There are plenty of hotels in Gaziantep, but none in Yeşilce. It is also easy to work the area from Birecik. There are several restaurants along the roadside near Yeşilce, but take some provisions along with you since a long walk may be required to find all the species.

Strategy

The best time of year is from May to September, but for Pale Rock Sparrow, Cretzschmar's Bunting and Cinereous Bunting, a spring visit is recommended. Little is known about species present in winter.

The area is easily worked in a day, but anyone visiting the area could usefully spend any spare time looking for new sites, since all the species should occur throughout these hills. The best time of day is early morning, but it is surprisingly easy to find the birds throughout the day, the only obstacle being the heat.

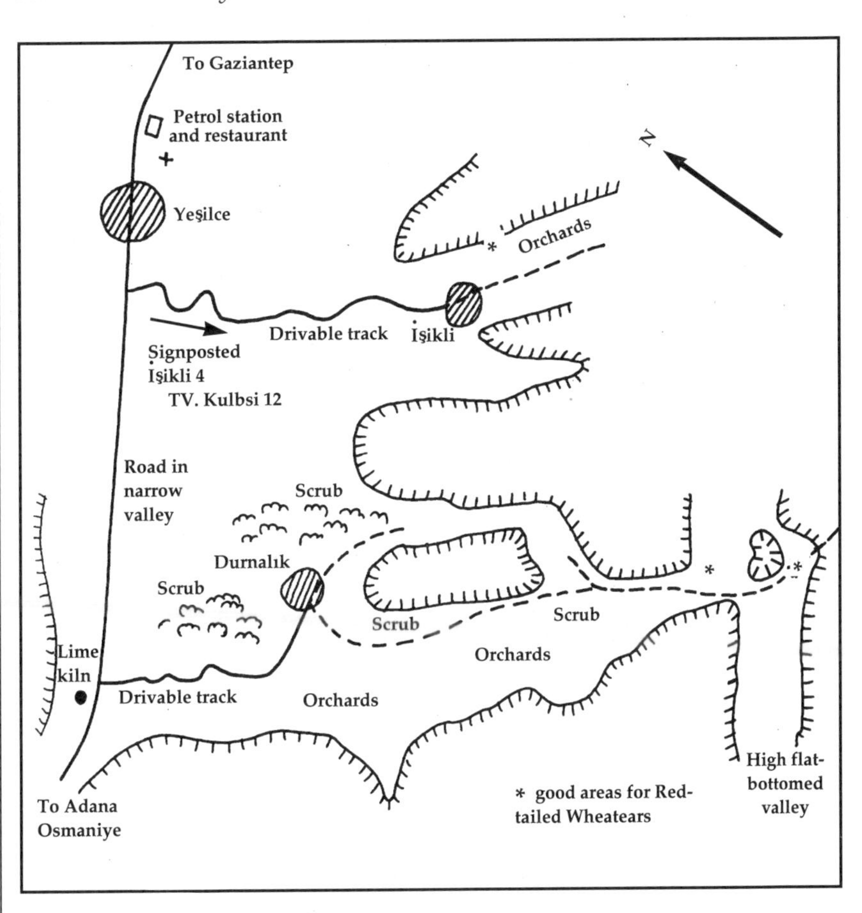

Birds During the summer, the valleys are packed with a truly amazing number of birds. Chukar are more usually heard than seen on the crags, while both Rock and Eastern Rock Nuthatches are easily located by the amount of noise they make. In the orchards and scrub, Orphean, Olivaceous and Upcher's Warblers can be found. Desert Finch, Black-headed, Cretzschmar's and Cinereous Buntings are present on the scrubby hillsides and the less well vegetated slopes have Pale Rock Sparrow. The latter are not recorded every year, and are best located by their buzzing call, but can be difficult to see. Red-tailed Wheatears are found at the far ends of both valleys on the barren slopes and Red-rumped Swallows can be seen hawking over the fields and streams.

İdil & Cizre

These two towns along the Syrian border provide opportunities to see some more elusive species. At the time of writing the region is very sensitive as it is not far from the Iraq border and a visit to the area is not advisable. Check the current situation with the Foreign Office before venturing into the area. Try to inform the military in the area of your intentions - they have even been known to house birders to keep them safe, but be prepared to be refused entry into the region. Do not stray from the roadside at Cizre, as it will lead to almost certain arrest. The

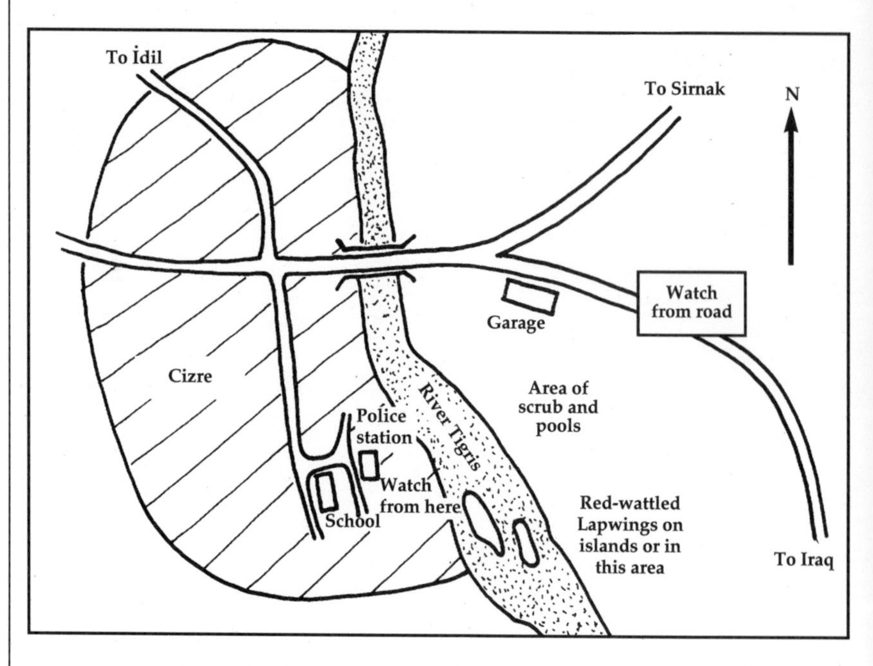

two important species in the area are Pale Rock Sparrow (one of only a few sites) and Red-wattled Lapwing (the sole site).

Cizre can be reached from the main tanker route to Iraq from Mardin (the E90) or from Bitlis via Siirt and Şirnak. İdil is 30km west of Cizre on the Mıdyat road. At İdil, there is a wadi that crosses the Cizre road about 6km east out of town. Look around this area for the Pale Rock Sparrows. Also in the area are See-see Partridge, Rufous Bush Robin, Finsch's Wheatear and Spanish Sparrow. At Cizre, the important species are all found along the Tigris. Look either from the bridge or the back of the garage on the Iraq road about 2km east of town (don't leave the road). Red-wattled Lapwings are on the gravelly banks, Black Francolin frequent the scrub and Pied Kingfishers hunt over the river.

There is little in the way of accommodation in the area, and it is best to get in and out of the area in a day unless you are enjoying the hospitality of the army.

Southern Coastlands

The Southern Coastland region is dominated by the Toros mountains. This range runs adjacent to the sea along most of its length, with spectacular cliffs marking the coastline. The mountains themselves have steep-sided gorges and high plateaux. The few low lying coastal areas provide habitat for several interestingspecies, both resident and migratory. Typical birds of the region include Black-eared and Finsch's Wheatears, and Lesser Grey and Red-backed Shrikes. The raptors most commonly seen in the higher areas are Griffon Vulture, Booted Eagle and Peregrine.

Göksu Delta

This is a large flat area where the Göksu river enters the Mediterranean Sea, and includes two lakes, Akgöl and Paradeniz, which are havens for birds and have recently been designated as nature reserves. Akgöl is fringed by tall reeds, whereas Paradeniz is virtually devoid of vegetation. A huge sandspit reaches out into the sea south of the lakes, while the area to the north is farmland.

Location

The lakes are the best areas for birding and are located south of the town of Silifke. To get to them, take the road from Silifke towards Antalya, and just before the port of Taşucu, look for a large fenced-off factory on the left. Take the road along the eastern side of the factory and just before the sea, cross the stream and turn right. The road continues along the coast past a holiday village, and eventually comes to an old gravel airstrip. Cross the strip and follow the tracks south-eastwards. There are several routes across the sands, but the reeds at the edge of Akgöl can be seen from the airstrip, so keep these in view. A track can be followed round to the eastern lakeside (where viewing is easier) until it reaches a dyke which joins the two lakes. A stony road will take you from the airstrip to a promontory on the eastern side of the peninsula, which can be good for seawatching.

Key

Sand dunes ⊙

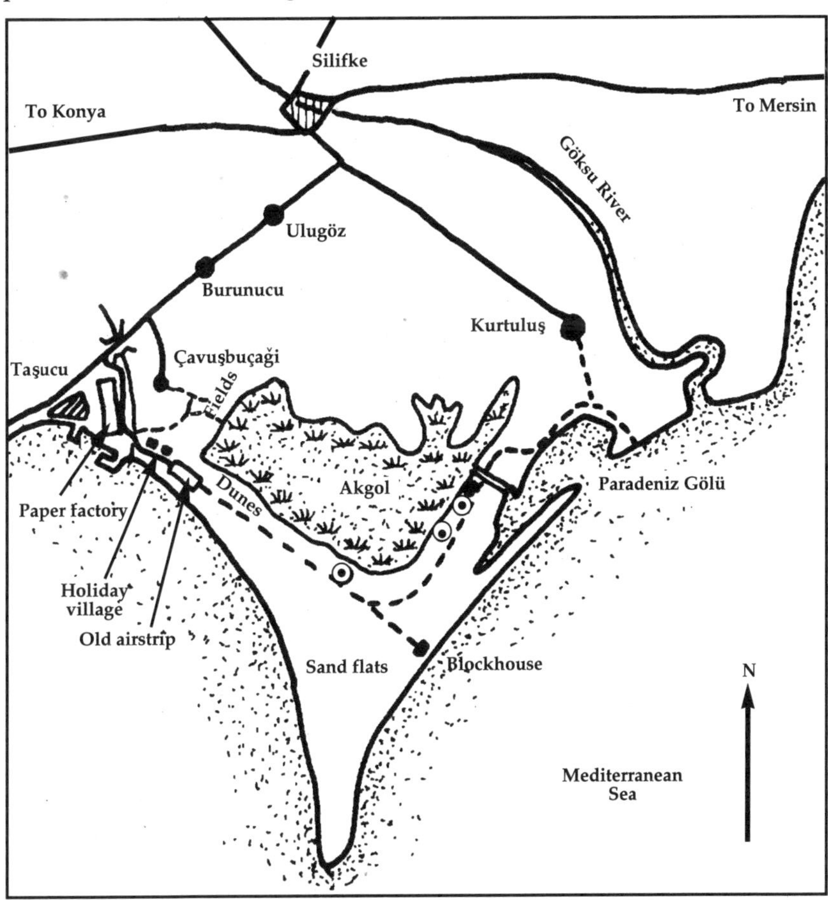

The paddyfields on the northern fringe of Akgöl can be reached by taking the track around the back of the holiday village.

It is not easy to reach the delta by public transport. It is possible to catch a bus or taxi to the factory from Silifke or Taşucu, but from there it is still a long walk to the lakes. It may be possible to hitch from the main road to the holiday village, but there is little traffic. There are also ferries to Cyprus (Kıbrıs) from Taşucu.

Accommodation

There is plenty of accommodation in the area, although the coastal areas may be full up in the summer. The holiday village has flats, and Taşucu has a number of pensions. The Hotel Çadır in Silifke is reasonably priced and has good facilities including an excellent restaurant (and quite often a Turkish Wedding which you can watch). There are also several campsites between Silifke and Adana with varying facilities, and there seem to be no restrictions on camping or sleeping out in the dunes at the delta.

Strategy

The best time to visit the site is during spring or autumn migration, when the resident birds are joined by thousands of migrants. In late April to May, the Black Francolins will be calling and easier to locate. During the summer, it gets very humid during the day and it is therefore advisable to visit in the early morning or the evening. In winter, the climate is warm and there are notable numbers of wintering wildfowl and some passerines.

Akgöl is surrounded by tall reeds making viewing difficult from most areas. However, along the southern shore, several dunes give reasonable views over the open water, and on the eastern shore there are some open marshy areas and pools. Paradeniz is open and easily viewable. The paddyfields to the north of Akgöl are particularly good for herons and egrets.

Birds

Resident birds include Pygmy Cormorant, Little and Great White Egrets, Purple Gallinule, Marbled and Ferruginous Ducks, Ruddy Shelduck, Black Francolin, Spur-winged Plover, Crested Lark and Graceful Warbler. All these species are reasonably easy to find with the exception of Purple Gallinule, which requires some searching around the edge of Akgöl, and Black Francolin which occurs in the surrounding scrub, but is elusive. To find the Francolins, it is best to listen for males in spring when they usually call from the top of a small bush or dune. Try not to flush these calling birds, since they are under pressure from illegal hunting and should not be disturbed. Peregrines and Marsh Harriers are regularly seen hunting in the area.

In spring and autumn, passage waders are present, but are difficult to see due to the reeds. Greater Sand Plover and White-tailed Plover have been recorded among the commoner species, and Turkey's only record of Crab Plover comes from this site. Some waders also frequent the seashore away from the holiday village. Other migrants include White Pelican, Glossy Ibis, Little, Baillon's and Spotted Crakes, and Montagu's and Pallid Harriers.

During the summer, Little Bittern, Purple Heron, Water Rail, Black-winged Stilt, Collared Pratincole, Rufous Bush Robin and Reed Warbler can all be found. In winter, the area is a haven for several species of duck and a few Spotted Eagles.

Seawatching off the delta can produce Shag and Audouin's Gull. In summer, and on passage, Cory's Shearwater, and Gull-billed and Common Terns can be seen. Arctic, Long-tailed and Pomarine Skuas, and Caspian and Lesser Crested Terns have all been seen from here.

The hillsides north of Silifke towards Mut and the fields by the Adana to Antalya road hold Yellow-vented Bulbul, and the pier at Taşucu has roosting Audouin's Gulls.

Other wildlife

The vast sandy beaches of the Göksu Delta are the breeding grounds for both Loggerhead and Green Turtles. Both these turtles are rare species in the Mediterranean and so disturbance should be avoided, especially at night when they are nesting. However, it is possible to see their tractor-like tracks where they have emerged from, or returned to the sea. It is possible to distinguish between the two species' tracks - Green Turtles move both their front limbs simultaneously to propel them over the sand while Loggerheads use front left and back right, followed by front right and back left. The dunes are home to a variety of reptiles - lizards and tortoises are the commonest but if you are lucky you may find a Chameleon.

Demirkazık

The mountain of Demirkazık is in the Aladağ range. The mountains are a mixture of grassy terraces and crocky crags and provide spectacular scenery and unusual birds.

Location

The village of Demirkazık is about 6km north of the village of Çamardı, and can be reached by following the signs to the mountain centre about 2km north of a petrol station. Cross the river and follow the road through the village and up to the mountain lodge. From the mountain lodge, the only way up is to walk, and it is quite strenuous. The alpine terrace above the lodge contains most of the area's specialities. See strategy for details.

Çamardı can be reached by bus from Pozantı, and Demirkazık by taxi or hitching from Çamardı.

Accommodation

Until recently, there was only one place to stay at Demirkazık - the mountain lodge, but from 1991, there has also been a pension nearby. The mountain lodge has either small rooms or dormitory accommodation, and prices seem to vary depending on how rich you look. There is usually no food available, unless there is a group staying, and if this is the case, it may be full, but this is not common during summer. You can also camp in the grounds. The pension is run by Ali Şafak, and is on the left a kilometre north of the aforementioned petrol station (it is the only two-storey building). Ali is a guide in the

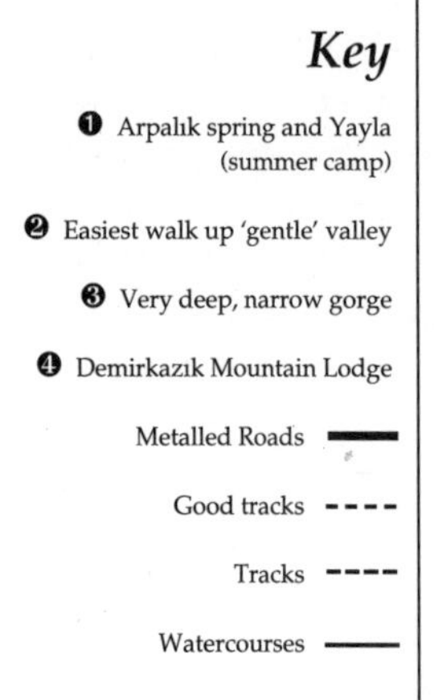

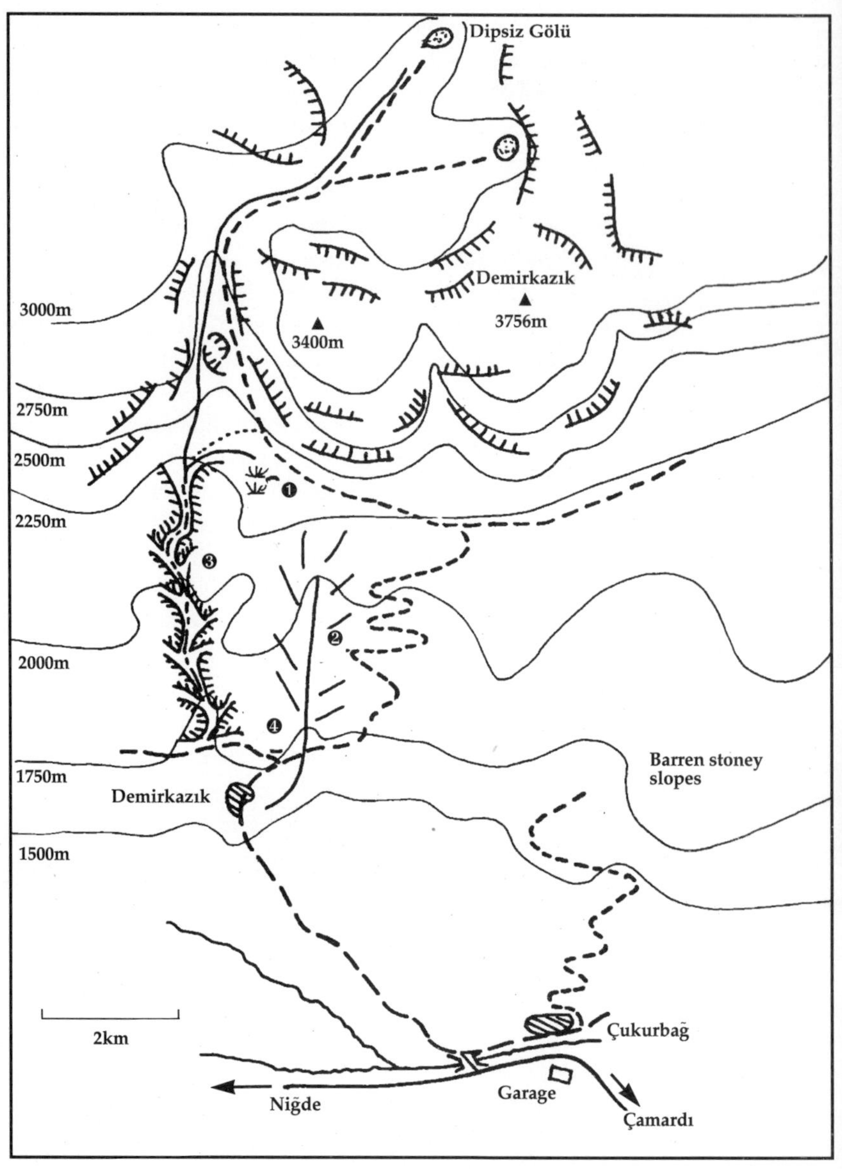

mountains in summer and you can hire him and his equipment (he has tents and donkeys to carry gear) if you want to explore the mountains. He is a genial character, with a weakness for Turkish folk music and rakı. The only nearby restaurants are in Çamardı and a good, cheap one on the Tarsus to Ulukışla road next to the Çamardı junction. You will also require provisions to take the mountain so buy these in Çamardı or elsewhere.

Strategy

The best time to visit the area is from April to May when there is still some snow on the ground, but the temperature is not freezing. Later visits will secure most species, but the Snowcock will be difficult to see. Winter is only for the brave and well equipped.

All of the species present can be seen in one morning, but it requires an early start. The best plan is to leave about one hour before sunrise and walk up the small gorge behind the mountain lodge. Take a torch, since there are a couple of barbed wire fences to cross. This will take at least 45 minutes, and will bring you out onto the terrace. Search the area up to the crags, and listen for the curlew-like call of the Snowcock. They climb into the crags soon after dawn and the only way to find them then is to scan the crags or to climb up into them, which is both difficult and dangerous. After this, search the hollow to the north and either descend by the original gorge, or the slightly more difficult one at the bottom of the hollow which comes out just north of the lodge. This second gorge is absolutely spectacular and looks impossible to descend in some areas, but is less difficult than it looks.

Birds

The bird of the area is Caspian Snowcock. See above for tips on location. On the lower slopes, Chukar, Ortolan and Rock Bunting are common, and Chough and Alpine Chough feed in the fields. On the terraces, Shore Lark, Tawny Pipit, Isabelline, Finsch's, Black-eared and Northern Wheatears and Snow Finch occur. The rocks hold Rock and Blue Rock Thrushes, Black Redstart, Alpine Accentor and an occasional Crimson-winged Finch. The hollow above the steep gorge has Radde's Accentor, and the top of the gorge is good for Wallcreeper, Red-fronted Serin and Crag Martin. The most commonly recorded raptors are Griffon Vulture, Lammergeier and Golden Eagle.

Radde's Accentor

Other wildlife

The most obvious feature of the limestone mountains of the Aladağlar throughout the spring and summer is the bewildering array of flowers. What looks like a bare slope will, on closer inspection, contain a seemingly different species at every step. This is particularly true of the *Astragalus* group of vetches with a great many of Turkey's 370 species to be found here. Small pockets of the beautiful Cedar of Lebanon can still be found in this area. Butterflies are also a noticeable feature of the alpine slopes and a day here should provide the chance of seeing Apollo, False Apollo and a variety of blues, coppers and fritillaries. In the high crags above the mountain lodge you may see

Ibex and Chamois, although they are both difficult to approach. On the lower slopes there may be Fox.

Nur Dağları

This mountain range separates the Seyhan/Ceyhan lowlands from the arid hills of southeastern Turkey. They provide a corridor for migrating raptors, and autumn counts can be impressive. The two known watchpoints are Belen and Toprakkale, although there are doubtless many more.

Sertavul

The village of Sertavul is 40km north of Mut on the road to Karaman. The cliffs above the village are good for raptors, notably Griffon and Egyptian Vultures, Black Kite, and Short-toed and Booted Eagles. The rubbish tip in the valley visible from the petrol station is also good for the Kites and Egyptian Vultures. Ravens wheel among the raptors, and the hillsides hold Sombre Tit and Krüper's Nuthatch.

Akseki

Akseki is a small village in the Toros mountains on the road between Beyşehir and the coastal town of Manaugat, and is known as a site for White-backed Woodpecker and Olive-tree Warbler. Both occur in the area, but are very difficult to find, and not many birders see them. The Olive-tree Warblers can be found either in an area of scattered trees behind a conifer plantation about 8km south of Akseki or at the graveyard in the southeast corner of the village. The former site holds Masked Shrike and occasionally Great Spotted Cuckoo. To look for the woodpeckers, find the layby 8km north of the village on the right. Across the road from this is a clearing where they occur, although they could just as easily be found anywhere in the area. Other birds include Syrian Woodpecker and Krüper's Nuthatch and, in summer, Orphean and Rüppell's Warblers and Cretzschmar's Bunting.

The pine forests around Akseki contain a good population of the pink-flowered Kurdish Helleborine, and the extraordinary flowers of the Komper's Orchid *Comperia comperiana* can be seen in the cemetery in late May. Keep a look out for one of the more distinctive butterflies which is not in the European field guides - *Thaleropis ionia* - a small admiral-type butterfly with intricately patterned upperwings and underwings, reminiscent of a Lesser Purple Emperor.

Black Sea Coastlands

The Pontic Alps form a narrow but substantial mountain barrier along the Black Sea coast. At the easternmost end of the chain, the northern slopes have a warm but very damp climate. South of the

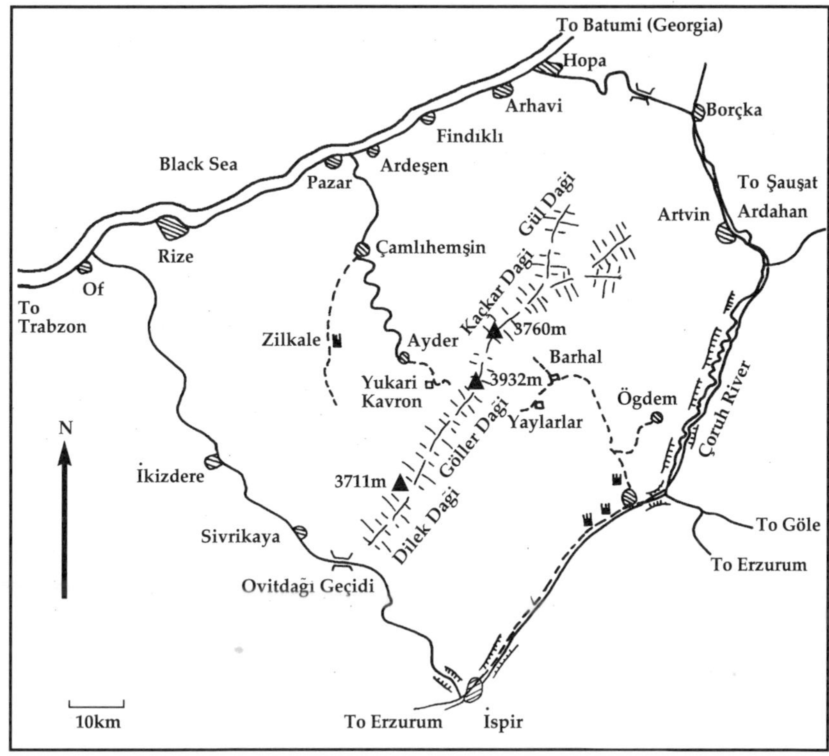

watershed, the weather is broadly similar to that of eastern Anatolia. As a consequence, the northern slopes are well vegetated, with tea plantations and hazelnut groves at low altitude and mixed and coniferous forests higher up, while the southern side is more sparsely vegetated. Further west, the mountains become less severe, with rolling hills and much agriculture.

Sivrikaya

This small village has become something of a Mecca for birdwatchers visiting Turkey, as it provides the best chance of finding Caucasian Black Grouse in the Western Palearctic outside of the Caucasus. The mountains rise to 3711m just to the north of Sivrikaya, and there is plenty of snow on the high crags, even in summer. Unfortunately, overgrazing has reduced the extent of the rhododendron, the favored haunt of Black Grouse, to a few steep north facing slopes. The only wooded areas to be found in the immediate vicinity are around the village itself, although there is plenty of forest further down the road towards Ikizdere.

Location

The village of Sivrikaya is a few kilometres north of the Ovitdağı Geçidi (pass) on the Rize to Ispir road. The road is usually passable, but landslides are common and can block the road, as can snow, even as late as May. There are signs of an improvement in the road; some widening and stabilising work is being undertaken between Sivrikaya and the top of the pass.

Key

Stream
Road
Crags
Ridgelines

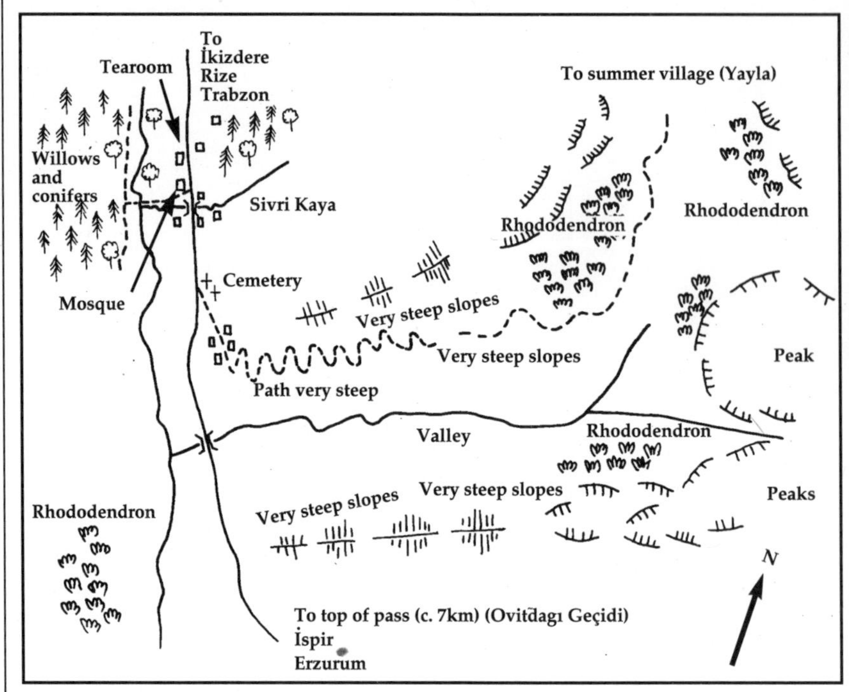

Accommodation

The only place to stay in Sivrikaya is the tea house, just to the north of the Mosque. It cannot be guaranteed to be open, especially in late summer, in which case the best option is to sleep in the open, even though the nights are cold at this altitude. There are hotels in Ispir and İkizdere (both are at least an hour's drive away), but to have the best chance of seeing Caucasian Black Grouse, an early start is necessary. There is usually an evening meal of some sort available in the tea house, although, if a late arrival is anticipated, it is best to eat in Ispir or İkizdere. There is a local shop where basic foods are available.

Strategy

The best months at Sivrikaya are late May and June, particularly for the grouse species. The weather can be a problem, even in the summer months with mists, rain and sudden storms being frequent. Whiteouts from fog are common until August, although the mornings are usually clear. The best chance of seeing the grouse is to get up among the rhododendron scrub shortly after dawn. It is a good two-hour walk to reach these areas from the village, so it is preferable to start in the dark. A favoured method, and normally a successful one, is to enlist the services of Mustafa Sari who lives in the village on the hill opposite the tea house. He charges up to £15 per person, but he does know where to find the birds. He may be up in the summer villages later in the year, in which case you will have to find the birds on your own. This is not too difficult, although finding the paths in the dark can be!

The main path starts a few hundred metres to the south of the tea house; head southeastwards past a graveyard, and then up the steep slopes. Look around the rhododendron scrub on the slopes for the grouse. For Caspian Snowcock, it is necessary to go up higher - the birds follow the snowline up into the crags as the year goes on. In September, the only realistic chance is to go right up towards the high peak, which can be seen at the head of the right-hand valley. Caucasian Black Grouse have also been seen in the rhododendron scrub on the side of the road, just south of Sivrikaya.

Caucasian Black Grouse

Birds

Caucasian Black Grouse and Caspian Snowcock should be seen here, although neither are common. They probably both occur in nearby areas of the Pontic Alps in suitable habitat, although road access is very limited in the region. A study in 1993 yielded good numbers of lekking Caucasian Black Grouse around Sivrikaya on both sides of the road.

Other high altitude species found on the upper slopes, particularly near the watercourses at the heads of the valleys, include Shore Lark, Water Pipit (very common), Alpine Accentor, Black Redstart, Ring Ouzel, Snow Finch, Twite and Crimson-winged Finch. Wallcreepers may be difficult to find, although the craggy peaks give plenty of ideal habitat. Scanning a cliff face from several hundred metres rarely brings quick results with this species but a good place to look, particularly in late summer, is on the road cuttings to the south of Sivrikaya towards the top of the pass.

Some good species can be found in the valley north of the village. Mountain Chiffchaff, Marsh Warbler, Dunnock, Wren and Bullfinch can be found in the scrub along the river, with Grey Wagtail and Dipper along the watercourse. Common Rosefinches are common all over the valley bottom, with smaller numbers of Red-fronted Serin and Rock Bunting. Green Warbler is best looked for in the scrubby fringes of the conifers on the far side of the river. Raptors are surprisingly scarce in the area - the road south to İspir is better, with a reasonable chance of seeing Lammergeier.

Other wildlife

Chamois and Ibex can be found in the high crags and meadows to the east of the pass and both Wolf and Bear could be seen in the area. Both of the latter are commoner to the north of Sivrikaya; indeed Wolves are seen quite frequently around Uzungöl and is presumably not uncommon in the forests above İkizdere.

In late July and early August look out for the very localised Yellow Globe Orchid *Traunsteinera sphaerica* in the alpine meadows on the slopes to the east of the pass. Butterflies include Apollo, Russian Heath, Pontic and Chelmos Blues and an isolated population of Gavarnie Blue. Coppers are well represented with Scarce, Fiery, Purple-shot and Sooty Coppers all found around Sivrikaya, and further down the valley towards İspir, Lesser Fiery Copper can be seen.

Uzungöl

Uzüngöl is a delightful village set at the head of a lake in a steep-sided valley on the northern slopes of the Doğu Karadeniz Dagları. It can be reached by road south from Of on the pass to Bayburt, and is visited by tourist dolmuses because of the setting and the mosque. This is another area where Caucasian Black Grouse can be found. There is also plenty of forest which should be worth exploring. The best way to look for the grouse is to walk up the valley side to the village of Yaylaronu, about two hours distant. This village is on the treeline, and a few grouse can be found at leks here in the late spring. They frequent the areas of dwarf birch/spruce scrub and rhododendron.

İspir

İspir is often visited by birdwatchers en route to or from Sivrikaya. However, it does provide good birdwatching in its own right, particularly as there is a good chance of seeing Semi-collared Flycatcher, a species which can otherwise be difficult to find.

Location

İspir is situated on the Çoruh Nehri (river), at the bottom of the pass going north to Sivrikaya. The town is normally approached on route 925 from Erzurum in the south, or from Rize in the north. The road eastwards along the Çoruh valley is of poor quality, but usually passable and well worth investigating. It starts just past the bridge on the western edge of İspir. There may be a 'Road Closed' sign, as the bridges are often washed away or damaged, but it seems that the sign is more or less permanent, so check with locals to find out if it is passable.

Bus services are reasonable to and from both Rize and Erzurum, although be prepared for delays on the former route (one bus driver is known to have driven into a river because his tyres caught fire on the Sivrikaya route!). There are dolmuşes on the Çoruh valley route, however hitching may be more effective.

Accommodation There are a few rather basic hotels and several rather poor restaurants in İspir. The town is big enough to support a range of banks, food shops and back-street bars.

Strategy If time is short, and Semi-collared Flycatcher is the priority, then the best areas to check are the parks in İspir itself, or the woods and orchards along the river between the town and the Rize/Erzurum junction. If more time is available, the Çoruh valley to the east is excellent, with a good raptor gorge about 6km from İspir, and on the slow trip through to Yusufeli, there are many small villages with orchards, woods, more gorges, and some very unusual multicoloured 'badlands' type scenery. There is another gorge on the road to Sivrikaya, a few kilometres to the north of its junction with the main road.

Birds In contrast to the high peaks north of İspir, the gorges around İspir are excellent for raptors. Griffon, Black (probably) and Egyptian Vultures and Lammergeier all breed in the area, the gorge along the Çoruh valley is particularly good for the latter species. Golden, Imperial, Steppe and Booted Eagles, Sparrowhawk, Buzzard, Montagu's Harrier and Hobby have all been recorded. The gorges also hold Alpine Swift, Crag Martin, Rock Thrush and Blue Rock Thrush. Wallcreeper has also been found here. Black Stork has been seen in summer and may breed in the area.

Aside from the parks and woods around İspir, the wooded areas all along the Çoruh valley to Yusufeli are a good place to look for Semi-collared Flycatcher. Golden Orioles and Syrian Woodpeckers are both quite common, while on autumn migration, Thrush Nightingale, many warblers, Spotted Flycatcher and Ortolan Bunting should be seen.

Sumela Monastery

Sumela monastery was founded in Byzantine times, and is well worth visiting to see the remarkable building which has been cut out of a cliff face, as well as for the local birding. As altitude is gained southwards from Trabzon, the tea and hazelnut crops on the terraced hillsides give way to rich, damp forest in the valleys. Visiting Sumela, after travelling through the rest of Turkey, is like suddenly being transported to the Pyrenees or the Alps, with its rushing streams, damp woodland and cool temperatures.

Location

To get to the monastery, take route 885 (E97) from Trabzon to Maçka. Just prior to Maçka take the left turn signposted to Meryem Ana Monastırı (It doesn't mention Sumela on the sign). The monastery is 25km up this road. A parking area, and some tourist facilities are located by the road, and it is here that the first, initially rather disappointing, views of the monastery are obtained - it's a lot better closer up!. The road up from Maçka follows a narrow valley containing a delightful rushing stream, however, this idyllic babble can easily turn into a raging torrent in a matter of hours, washing away parts of the road quite regularly.

Accommodation

Maçka has the nearest hotel accommodation, although Trabzon is not too far away and offers a greater range, mostly in the Taksım area of town which is easily found from the airport. Follow the signposts for the town centre (Şehir Merkezi/Centrum), and you will soon be among them. Regular Turkish Airline flights serve Trabzon from a number of main cities. There is a ferry service from İstanbul, calling at Samsun, and taking two days. There are plenty of buses and dolmuşes from Trabzon to Maçka, and from both towns there are dolmuşes taking tourists to the monastery. Hire cars are available from Avis in Trabzon.

Strategy

From the car park, follow the path past the tourist café, climbing through a series of switchbacks to the monastery 250m above. From the monastery, a good path follows the valley southward, slowly losing altitude until it reaches a track which is a continuation of the road to the car park. This circuit can be done in two or three hours. Green Warblers can be found throughout the circuit, although they are more common on the way up to the monastery. Conditions are often dank and misty, even in the summer months, and particularly in the afternoons. Green Warblers can be more difficult to find in the summer, although they are present until at least the beginning of September.

Birds

Green Warblers are probably found much of the way along the Pontic Alps, even as far west as Bolu, however the lack of coverage of much of the area means that they have only been recorded from a few sites. At Sumela they are fairly easy to locate, especially in spring, when their wagtail-like call is helpful in finding them.

Common breeding species include; Great Spotted Woodpecker, Grey Wagtail, Dipper, Crag Martin, Treecreeper, Marsh and Coal Tits, Goldcrest, Blackbird, Robin, Jay, Raven and Crossbill. In this area, Cirl Bunting reaches its easternmost point in Turkey.

In recent years Red-breasted Flycatcher has been found breeding in the woods around Sumela, and is in fact quite common in the valley below the monastery. Black Woodpecker has been recorded, and may well occur all along the northern side of the eastern Pontics. Tengmalm's Owl was heard below Sumela in 1990 - records of this species are very sparse in Turkey but it is quite possible that they breed in the area.

Other wildlife

Sumela is surrounded by an excellent stand of euxine forest, the climax vegetation of these north facing slopes. *Pinus orientalis, Abies nordmanniana*, and the stately Oriental Beech are some of the dominant trees. If you visit in the early summer the rich pink flowers of *Rhododendron ponticum* put on quite a show and among them there will be smaller numbers of *R. luteum* with rich yellow flowers. Honey made with pollen from this species is poisonous and apparently caused many soldiers to go mad when Xenophon and the ten thousand made the long march over these mountains. Wild Boar are common in the forest.

Kızılırmak Delta

The wide flat delta at the mouth of the Kızılırmak River provides plenty of good habitat for both breeding and wintering species.

Location

The delta is situated to the north of the main Black Sea coastal road (R 010) near Bafra, and is easily reached from Samsun which has regular air, boat and bus services from İstanbul. There are also boat and bus services from Trabzon.

Accommodation

There are a few standard hotels in Bafra, and a wider range in Samsun, 50km distant.

Strategy

The area can be rewarding throughout the year. It has a number of interesting breeding species, good populations of wintering birds, and it acts as a magnet to birds migrating over the Black Sea. The most productive area is around Balık Gölü and Uzungöl. This is reached

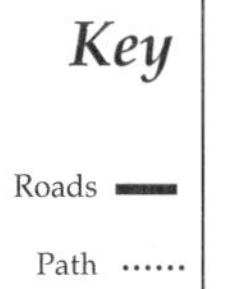

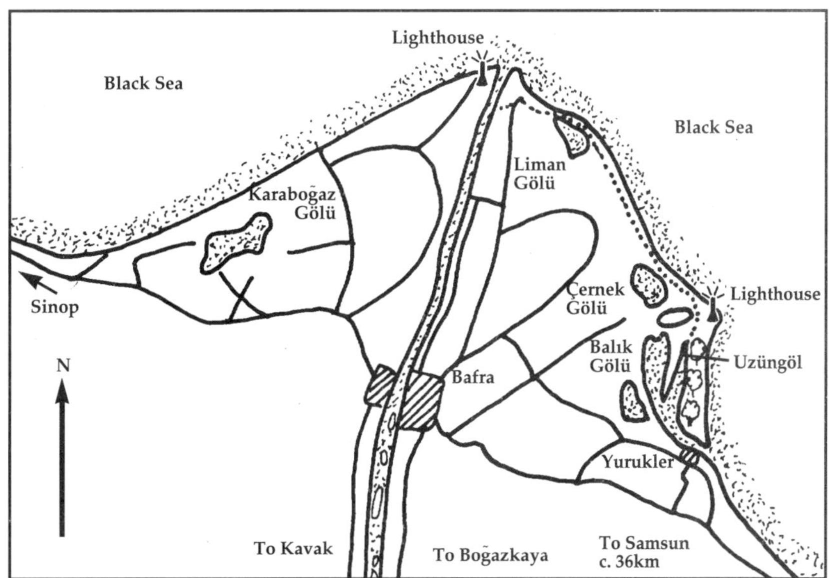

from the Samsun to Bafra road by turning north at Engiz towards Yürübler. Here a bridge crosses a canal, and a walk along this can produce several species. Across the canal the woods are worth exploring for woodpeckers. West of Yürübler, after crossing two canals, there is a track northwards to a fishing station (Balıkhane). Good views over Balık Gölü and Uzungöl can be had from here.

The other large lake, Karaboğaz Gölü can be difficult to view, because of large flats surrounding the lake, but can be reached north off the Bafra to Sinop road, about 15km. west of Bafra.

Birds

During winter the lakes hold good populations of wildfowl, with up to 25,000 ducks and geese present. The commonest are Shoveler, Teal and Wigeon, but White-headed Duck, Red-breasted Merganser, Goosander and Smew can all be found. There are also several Spotted Eagles which usually winter here.

The woods near Yürükler have Green, Great Spotted and Lesser Spotted Woodpeckers and an isolated breeding population of Wryneck. Redshank, Kentish Plover and Black-winged Stilt breed around the lakes, and Marsh Harrier and Bittern can both be found. Dalmatian Pelicans are regular, and White-headed Duck breed in small numbers. White Storks breed in the trees around Yürükler.

Other wildlife

Both Stripe-necked and European Pond Terrapins can be found in the delta, along with Common Tree Frog. The coastal dunes have a good population of *Pancratium maritinum*, a beautiful 'Sea Daffodil' which flowers in the autumn.

Kaçkar Dağları

The Kaçkar is the highest region of the Pontic Alps, culminating in Kaçkar Tepe at 3932m. The northern flanks are cloaked in thick forest, a mixture of pine, spruce and various hardwoods, up to about 2200m. The climate on this side is very damp. On the southern slopes the climate is much drier, though the valleys still contain forests, which are more open, and there are plenty of walnut, cherry and pear orchards. The alpine zone is extensive and snowbound well into summer. The passes over the main divide are often not open until August.

Location

The Kaçkar lies between Ardeşen on the Black Sea coast and Yusufeli in the Çoruh valley to the south.

Accommodation

There is a good quality hotel in Artvin some 60km east of Yusufeli, and a number of standard ones there and in Yusufeli. Yusufeli has geared itself up as the trailhead for trekking in the Kaçkar since the publications of 'Trekking in Turkey' by Enver Lucas and Marc Dubin (Lonely Planet Guides, 1989) and has a surprising amount of accommodation. Further into the range there are rooms to be found in Altıparmak (also known as Barhal) and Yaylalar (altitude 2200m), both

Key

❶ Çaymakçur Geçidi Pass at 3205m – Open from late July

❷ Korahmet Geçidi Pass c.3100m

❸ Dupeduz – Good capsite

❹ Dilber Düzü – Good campsite, pass above is very difficult

Metalled Roads

Good tracks

Tracks

Paths

Streams

Peaks ▲

Ridgelines

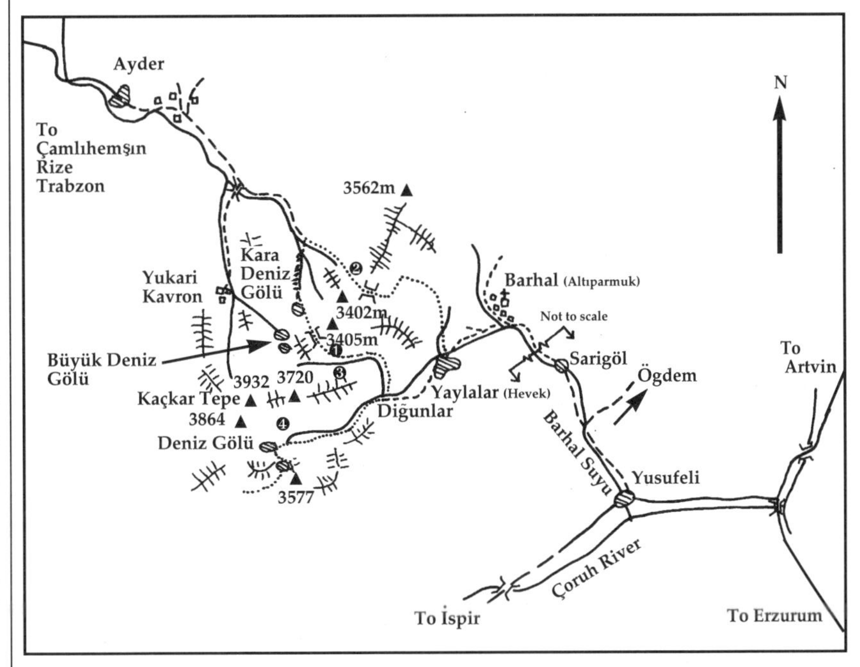

within walking distance of the high peaks. On the north side rooms are available at Ayder. Much further down on that side there is a 'fieldcentre' about 6km south of Çamlihemşin which can be recommended for exploring the wonderful forest in this area or as a base for travelling higher. The owners speak good English and play host to fieldtrips from Turkish universities. The eerie Zilkale (Georgian fortress) further up the same valley is straight out of fantasy land!

Strategy

The forests are rather hit and miss. Those on the northern slopes are probably richer, but the terrain is very difficult, and the undergrowth often very nearly impenetrable. The alpine species are best sought from the south side (with one exception). Birding in this zone is best attempted in high summer, as snow cover is extensive even as late as June. There is quite a lot of permanent snow above the 3000m line. Even though the 3000m line can be attained from both Yaylalar and Altıparmak in a day, this really is not an area where a dayhike to altitude would be advisable. A three-day trek with a guide and/or mules should give an adequate flavour of the area. Both are readily available from both villages and the prices are not high (c. £30 a day for mule with guide). The walking is as strenuous as you want to make it, the scenery is stupendous, and once altitude has been gained the birds aren't bad either! A recommended route for the species mentioned below is from Yaylalar: Day 1 - Yaylalar to Düpüdüz, where there is a good alpine meadow campsite; Day 2 - Düpedüz to Karagöl, camping by the lake at 2900m - cold at night; Day 3 - return to Yaylarlar - a long but downhill walk. When walking in the area a good map is indispensible. 'Trekking in Turkey' contains a reasonable map and good directions, and there is a very good map in existence which at the present time is not available to the public but hopefully will be soon.

Caspian Snowcock

Birds

The forested zone on the northern slopes abounds with Green Warblers, and Red-breasted Flycatchers are quite common. Mountain Chiffchaffs, Common Rosefinches and Red-fronted Serins can be found around the treeline. White-backed Woodpecker, Black Woodpecker and Tengmalm's Owl haven't as yet been seen in these forests, but must surely occur.

The high alpine zone contains relatively few species, but the populations of Shore Lark, Alpine Accentor and Wallcreeper are perhaps unequalled in any other Turkish mountain area. Water Pipits, Chough, Alpine Chough and Crimson-winged Finches are also quite plentiful. Lammergeier and Golden Eagle are the two most commonly observed raptors.

This must also be the best area in Turkey to find Caspian Snowcock. They are quite common on the ridges around the 3000m mark (in August), especially just over the divide on the northern side. Caucasian Black Grouse are much less common at altitude but can be found in the rhododendron scrub along the northern side of the divide, and on both sides in the Altıparmak Dağı, as they move up from the treeline in the high summer months. Early morning at Karagöl (see above), in August, is a cacophony of sound as Snowcocks call loudly from the ridges all around, along with a few Caucasian Black Grouse call; some of the birds from ridges only a couple of hundred metres away.

Other Wildlife

Wolf and Bear are not uncommon in the Kaçkar. Both should be looked for around the treeline, although Bears frequent the forests down to valley levels (the broken branches in many of the fruit orchards are a sign of their foraging) and the Wolves can be seen up to the highest altitudes. Chamois and Ibex are also present but have been quite heavily hunted. The floral display is fabulous, and officianados of orchids, primulas, louseworts, gentians, bellflowers and fritillaries probably won't have time to look for birds!

Borçka

The far northeastern end of the Pontic Alps is rugged with steep-sided valleys, high mountains and many fast-flowing streams. The area hosts an extraordinary raptor migration in autumn. The birds are concentrated along the Georgian coastal region and funnel south along the ridges into Turkey where they can be seen in huge numbers at various locations around Borçka, Arhavi and even further inland (see Çam Geçidi).

Location

Borçka is situated in the Çoruh valley about ten kilometres south of the Georgian border. Arhavi and Hopa are coastal towns which are connected to Borçka by the Cankurtaran Gecidi. Trabzon is about 150km to the west and has good ferry and air links. Hire cars are available in Trabzon.

Accommodation

The Karahan Otel (Inonu Cad. 16, Tel 1800) in Artvin is quite good, and there are reasonable hotels in Arhavi (Hotel Azur) and Hopa (Çihan Oteli, Ortahopa Cad. 7, Tel 1897). Borçka doesn't have much to offer, although accommodation can be found. There is quite an army presence in the area, particularly in Borçka and Hopa, so care should be taken especially if there is unrest in Georgia as the Turkish garrisons are strengthened and tend to be a little edgy on these occasions.

Strategy

Although birdwatching has plenty to recommend it from spring to autumn in this area, unquestionably it is the latter season that provides the main reason for visiting Borçka. Raptor migration has recieved little coverage in this region and a few records of raptors on migration in the sixties and early seventies did not prepare a team of British observers for what they would find in an autumn of coverage in 1976. In total, c. 370,000 raptors of 28 species were logged between 17th August and 10th October. Since then there has again been little proper study of the migration which is somewhat surprising given the huge numbers involved.

The raptors tend to follow two main routes, a coastal one through Arhavi and another aligned roughly with the Çoruh Valley. The former tends to be used by raptors passing early in the season, with fewer birds and fewer species. The weather can be very volatile at this time of year, so be prepared for often torrential and seemingly endless rain, and even snow at quite low levels in October. Temperatures vary considerably from cold, particularly on the hilltop watchpoints, to quite hot on the sunnier days. The exact effects of weather changes on raptor migration in the area are not clear. Any hilltop near Borçka could provide excellent birdwatching in the right conditions.

A good site can be found by crossing the river from the south in the town and where the road forks, taking the left gravel track. This will take you to a ridge after about 4km. Another is a hill reached by a track on the east side of the Çoruh (head south along the river), although the going can be difficult depending on the level of the water. A third good watchpoint is the hill southwest of the town, reached by taking the Göktas road and then the first track off this on the south side. The track crosses a bridge and then snakes up the hill.

Birds A breakdown of the raptor counts in the autumn of 1976 provides a good summary of numbers and species likely to be encountered at this season. 205,000 Buzzards *(B. b. vulpinus)*, 138,000 Honey Buzzards, 5,775 Black Kites, 736 Lesser Spotted Eagles, 688 Sparrowhawks, 290 Levant Sparrowhawks, 473 Booted Eagles, 271 Steppe Eagles, 243 Short-toed Eagles, 133 Pallid Harriers, 124 Montagu's Harriers, 385 Marsh Harriers, 189 Hobbies, plus smaller numbers of Egyptian and Griffon Vultures, Hen Harrier, Goshawk, Long-legged Buzzards, Spotted, Imperial, Golden and White-tailed Eagles, Osprey, Lesser Kestrel, Kestrel, Red-footed Falcon, Saker and Peregrine. A count conducted later in October showed that relative numbers of Griffon Vulture, Hen Harrier, Sparrowhawk, Goshawk, Steppe, Spotted and Imperial Eagles, Kestrel and Merlin increased at this end of the season (although it is possible, but unlikely, that this was a variation of the year). Spring migration is practically unknown as there have been few visits by ornithologists at that season. Of course, other birds do occur in the area and not just on migration! The forests are excellent, with a good cross section of the typical forest species of the eastern Pontics and, in particular, White-backed Woodpeckers can be found here. The river mouth at Hopa is also worth checking for migrants.

Çam Geçidi (Şavşat)

The Yalnızçam Dağları runs southwestwards from the Soviet border, part of a series of parallel mountain ridges. The western slopes are well wooded, with coniferous and mixed forests and many open meadows. The range is more rolling than mountainous on top, and bears a superficial resemblance to Scottish moorlands.

Location Çam Geçidi is the pass between Şavşat and Ardahan, the top of the pass being at 2640m. The whole area is worth a look, and the next pass to the south, the Yalnızçam Geçidi, should be worth checking, although the road may be difficult.

Accommodation There are hotels in both Ardahan and Şavşat. Şavşat, at 1000m, has some surprisingly good hotels for such a small place. Buses over the pass are infrequent, as is traffic in general. Hitching is probably OK as most vehicles will stop.

Strategy Between Şavşat and Ardahan there are a number of areas worth checking. Any areas of woodland on the way up hold woodpeckers. The meadow edges, and in particular the areas just above the tree line, are excellent for passerines. In autumn large numbers of raptors follow the east side of the ridge, and many can be seen in the valley below, towards Ardahan (see Ardahan Ovası and Çıdır Gölü, p78).

Birds

The woodlands on the way up to the pass have an avifauna more generally associated with western Europe; Great Spotted Woodpecker, Tree Pipit, Wren, Redstart, Blackbird, Mistle Thrush, Dunnock, Goldcrest and Bullfinch are reasonably common. Black Woodpecker is probably commoner than previously realised, and White-backed Woodpecker should be found locally. Green Warblers can be found in the many areas of mixed woodland on the slopes .

The tree line is a few hundred metres below the pass, and here Water Pipit, Alpine Accentor, Stonechat, Whinchat, Black Redstart, Rock Thrush, Snow Finch, Twite (the quite striking Turkish race *brevirostris*), and Red-fronted Serin can be found.

Where grazing hasn't been too intense (it has in most areas) there is still some rhododendron scrub on the high ground, and Caucasian Black Grouse have been seen here, although records are scarce. On the top, Northern Wheatears are common, and Pied Wheatears have been recorded. It is possible that they breed locally, but the records may just indicate stragglers from the nearby Soviet breeding grounds.

The raptor passage consists mainly of Steppe Buzzards, with smaller numbers of Honey Buzzard, Sparrowhawk, Long-legged Buzzard, Lesser Spotted, Spotted, Steppe and Booted Eagles, all the harriers, Hobby and Saker. There is a good viewpoint about 1km past the highest point of the pass, from which it is possible to see birds drifting along the ridge (some come by only a few metres away) and flying lower down in the valley. Raptors can be seen perched on poles and hayricks all across the valley floor towards Ardahan (see Ardahan Ovası and Çıdır Gölü, p78).

Other wildlife

The forests between Şavşat and Çam Geçidi hold a few Wolves, Bear and Fox. The thick forests halfway up the pass have a rich saprophytic ground flora including abundant Ghost Orchids as well as Bird's Nest Orchid and Yellow Bird's Nest. On the alpine meadows above the giant yellow-flowered *Centaurea macrocephalus* can be seen in flower in late summer along with a variety of Geranium, Delphinium and Aconitum.

Eastern Anatolia

Eastern Turkey is a complex area both geographically and ethnically. Much of the region is mountainous with many high peaks, including Turkey's highest, the spectacular conical peak of Büyükağrı Dağı(Ararat), rising to 5122m. The main vegetational types are essentially steppe and alpine, depending on altitude. Two major rivers rise in this region, the Firat Nehri (Euphrates) and the Aras Nehri (Arras), the latter forming part of the Armenian border. Their valleys are quite lush, although trees are rare. Agricultural practices such as haymaking are often carried out by hand although the current increase in mechanisation is likely to cause some decline in the characteristic breeding species of the valleys.

Van Gölü is Turkey's largest lake and, at 1720m, one of its highest. It was formed when Nemrut Dağı erupted and blocked the natural outflow of the basin. The main lake has a high mineral content and as such is relatively poor for waterfowl, although it is an excellent place to wash clothes - no soap needed! The lake is surrounded by mountains which form a beautiful backdrop to the lake's startling blue waters.

The area is snowbound for much of the winter, and even in April very cold weather is a possibility. From May to the beginning of October the weather is generally warm, dry and sunny. Nights can be cold and special care should be taken at high altitude. The summit of Suphan Dağı can be bitterly cold even in the daytime. As with any mountainous region the weather can change rapidly and violent storms often develop very quickly.

Towns such as Van and Doğubeyazıt are relatively used to tourism but in much of the rest of the region tourists are rarely encountered. The Kurds (and other local ethnic groups) are extremely kind and hospitable and problems with them are unlikely. Care should still be taken, as friction with the military, and the proximity of sensitive border areas, makes this a rather unstable region. Some roads are supposedly closed at night by the military.

Characteristic breeding birds of the area include Ruddy Shelduck, Egyptian Vulture, Long-legged Buzzard, Saker, Common Crane, Bee-eater, Roller, Bimaculated Lark, Tawny Pipit, eastern races of Black Redstart and Stonechat, Black-eared Wheatear, Finsch's Wheatear, Snowfinch, Crimson-winged Finch and Grey-necked Bunting. Migration is much in evidence, with waterfowl, waders, raptors and passerines all finding suitable stop-over points around the lake. The region's position at the very eastern fringe of the Western Palearctic makes for an ornithologically exciting area.

Van

The area around Van town offers some good birding opportunities. The marshes along the lakeshore and the hills behind the town provide a wide variety of habitats and there is much to be discovered in the region. The south marsh consists mainly of reedbed and *Scirpus/Carex* beds with a few areas of open water along the shore and wet grazing marsh behind. The northern marsh is much more open, but heavily disturbed, with buildings along most of the shore. The hills just to the north-east of Van are typical of much of the habitat to be found in the eastern half of this region, being basically high and barren. Characteristic high steppe and mountain species can be seen here, including the difficult and elusive Grey-necked Bunting.

Location

The town of Van provides an excellent base from which to explore the region, with the opposite lakeshore being only a three-hour drive away. The main route ringing the lake is very good with a lower percentage of potholes than perhaps any other main road in Turkey and very little traffic. Once off this route, the roads are usually unmetalled, and driving can become a difficult experience. Van is well

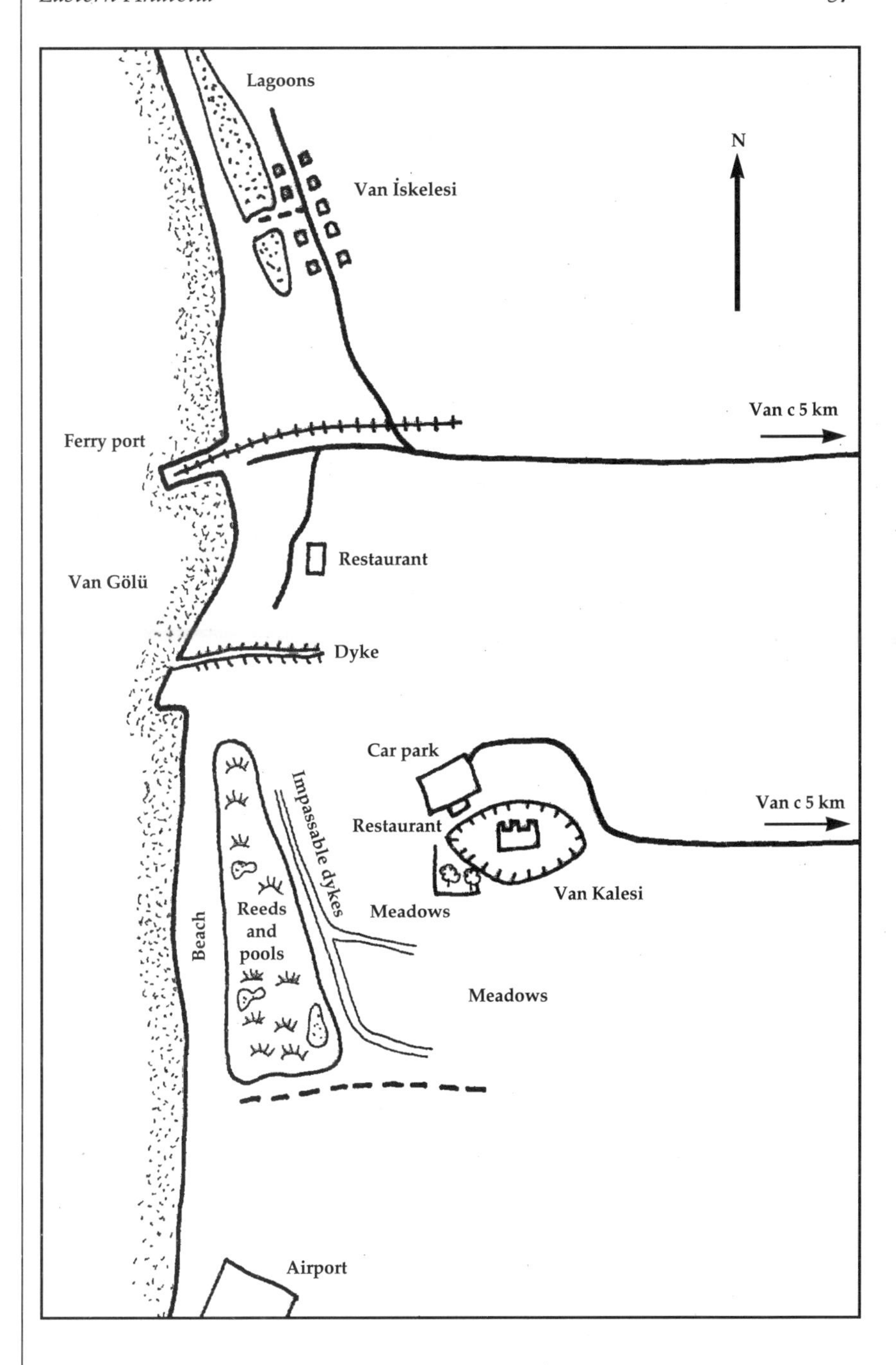

served by long-distance bus services and as usual there are plenty of local dolmuşes. It can also be reached by rail to Tatvan and then by ferry across the lake. There are regular air services from Ankara, Trabzon and Diyarbakır.

The marshes are situated on the lakeshore near to the ferryport and Van Kalesi (an ancient Urartu fortress built on an isolated rock about a kilometre from the shore). The easiest way to reach them is to catch a dolmuş or taxi (or drive) to the ferryport or castle depending on which part of the marshes you intend to visit. It is difficult to get to the shore without getting wet if you start from the castle. Dolmuşes go quite frequently to both the ferryport and Van Kalesi from the north end of Çumhuriyet Caddesi. If driving, the ferryport can be reached by heading towards the lake from the north end of Çumhuriyet Caddesi, crossing straight over the main Erciş road. Just before the ferryport there is a junction for the village of Van İskelesi to the right (this will take you to the north marsh), and a small track to the left leading to a restaurant/bar on the shore. From here a short wade across a dyke will take you onto the sandbar separating the marsh from Van Gölü. To reach the castle, turn left (south) on reaching the main road and take the next road right towards the lake. The hills at Van Toprakkale are about two hours walk from Van along the railway track, but if this method is used to get there, keep well clear of the track itself and don't use optics when trains are in sight as they often carry military personnel.

A good spot to find Grey-necked Buntings is by the edge of the road to Erçek. This road is signposted to Özalp and Iran off the main Erciş road just to the north of Van. It passes the main bus station and then rises steeply past a reservoir and a roadside rubbish tip until, just after the crest of the hill, the railway track crosses the road again. Park just before, or just after, the track and walk along the tops either side of the railway gorge south of the road.

Accommodation

Van is a friendly town with a wide variety of hotels, ranging from the quite plush Hotel Büyük Urartu and Hotel Kent, to a number of more basic etablishments situated on or near the Çumhuriyet Caddesi. At the cheaper end of the market the Hotel Van is quite reasonable and is situated near the post office on a small street west of and parallel to Çumhuriyet Caddesi. Avoid the Nuh Oteli - its a real misnomer! The hub of Van's social life is the large open tea garden on Çumhuriyet Caddesi. There are many good kebap houses and pastanesis in this area, and the Soydan restaurant on Maraş Caddesi is particularly good. For information on bus and dolmuş services, and for arranging taxis, the travel agent (between the Soydan Restaurant and Hotel Van) is helpful. It is possible to hire cars locally in Van, but at present, vehicle insurance is unobtainable.

Strategy

Van marshes can provide excellent birdwatching from the beginning of May through to October, although they are at their most interesting during migration periods. North Van Marsh is not really worth visiting. Only a few species are usually present, although these can include White-headed and Marbled Ducks and migrating waders congregate at the northern end. The Grey-necked Buntings at Van Toprakkale are best looked for in spring when they are singing, since they become elusive during summer when they sit tight during the day. In autumn, they begin to flock and can be found more easily again.

South Van Marsh can provide a real challenge. During May or September anything could pop up in front of you - although it will probably be tantalising glimpses of an odd looking *Acrocephalus* warbler! Most birdwatchers use the sandbar as a route to check out the marsh, however this is dependant on light conditions, with early mornings effectively useless as the sun is directly behind the reeds. Similarly, visiting the castle side should be done in the mornings, as evenings involve looking straight at the setting sun.

From the shore, a walk of about 2km along the sandbar will take you along the edge of the whole marsh. Herons, ducks, waders, terns and gulls tend to be seen in flight over the reeds or lakeshore, while passerines, particularly migrants, can be seen feeding along the edge of the reeds. The marsh peters out as it reaches the edge of the airport at the southern end. It is probably unwise to birdwatch in close proximity to the airport.

From the castle side access is restricted by a number of deep (too wide to jump) dykes, although an approach to the back of some of the reedbed is possible. This side is excellent for waders and herons. There is a small wood by the castle which is good for migrating passerines and the castle rock itself can also be interesting.

The hillsides are best worked by just wandering about. The Grey-necked Buntings can be located by their song in spring, but diligent searching may still be required to find them. Still, they do occur right down to the roadside on the Erçek road.

Birds

In the marshes most of the commoner breeding species can be seen from either side of the reedbed. These include Little Bittern, Night, Squacco and Purple Herons, Ruddy Shelduck, Ferruginous Duck, and the shyer White-headed and Marbled Ducks. Marsh Harrier and Hobby are usually present and probably breed locally. Black-winged Stilt and Kentish Plover breed on both marshes while Green, Wood and Common Sandpipers can be seen from May to October.

Of the reedbed passerines, Great Reed and Moustached Warblers are quite common, and the thick billed race of the Reed Bunting *(E.s. canetti)* occurs at the southern end of the marsh. Yellow Wagtails *(M.f.feldegg* breeding, with *M.f. beema* and occasionally *M.f. lutea* on

Crimson-winged Finch

migration) are common and Citrine Wagtails almost certainly breed here in small numbers. A recent discovery was of Turkey's first summering Paddyfield Warblers. Singing males were found here in 1986 by some Swedish birders and have been noted in most years since. The area around the castle has good numbers of Lesser Kestrel, Alpine Swift, Bee-eater, Roller and Hoopoe.

The nature of the marsh means that breeding proof is difficult to obtain. Red-necked Grebe, Egyptian Vulture, Little Tern, Bearded Tit, Penduline Tit and the '*armenicus*' race of Herring Gull are frequently found and may well breed. Common migrants in both seasons include Black-necked Grebe (mostly on the lake itself), Garganey, Avocet, Ruff, Snipe, Gull-billed, Whiskered and White-winged Black Terns, and, particularly in autumn, *Acrocephalus* warblers. White Pelican, Montagu's Harrier, Common Crane, Red-necked Phalarope, Spotted Crake, Sand Martin, Bluethroat and Golden Oriole are regular migrants. Rose-coloured Starlings have been recorded on a number of occasions, but their occurence is sporadic.

Grey-necked Buntings are presumably not an uncommon breeding species in the hills in this region, but they are rarely seen and relatively little is known about them. The Van Toprakkale site is the most commonly visited by birdwatchers looking for this species, but they should be found (especially in autumn when they start to flock) in similar habitat anywhere to the south and east of Van Gölü. Other species to be found here are Egyptian Vulture, Long-legged Buzzard, Golden Eagle, Black-eared Wheatear, Finsch's Wheatear, Rock Thrush and Crimson-winged Finch. Bimaculated Lark, Tawny Pipit and Stonechat are common on the lower slopes. Eagle Owls have been seen in the railway gorge and Pied Wheatear has recently bred near to the reservoir and are possibly widespread and under-recorded in the area.

Less common migrants and rarities which have been observed at the marshes include; Cattle Egret, Steppe and Spotted Eagles, Little and Baillon's Crakes, Whimbrel, Great Snipe, Black-winged Pratincole, Fan-tailed, Savi's and River Warblers and Iraq Babbler. A few days here in spring or autumn are almost certain to turn up something a little out of the ordinary.

Other wildlife

Van marshes are notable for their enormous numbers of frogs and terrapins, especially on the landward side of the wetland. Both Stripe-necked and European Pond Terrapins occur. In late May the meadows between the castle and the marshes are tinged purple in places from a dense population of *Orchis palustris*. There are usually a number of white albinistic forms there.

Erçek Gölü

Erçek Gölü is a large, and quite deep lake, at an altitude of 1890m. It is almost surrounded by high mountains, with some particularly craggy ones to the south (Baz Tepe). There is a flat valley floor to the east, which is seasonally flooded. The village of Erçek on the edge of the lake

Key

Metalled road

Driveable track

watercourse

Watercourse intermittant

Grazing marsh

Reedbed

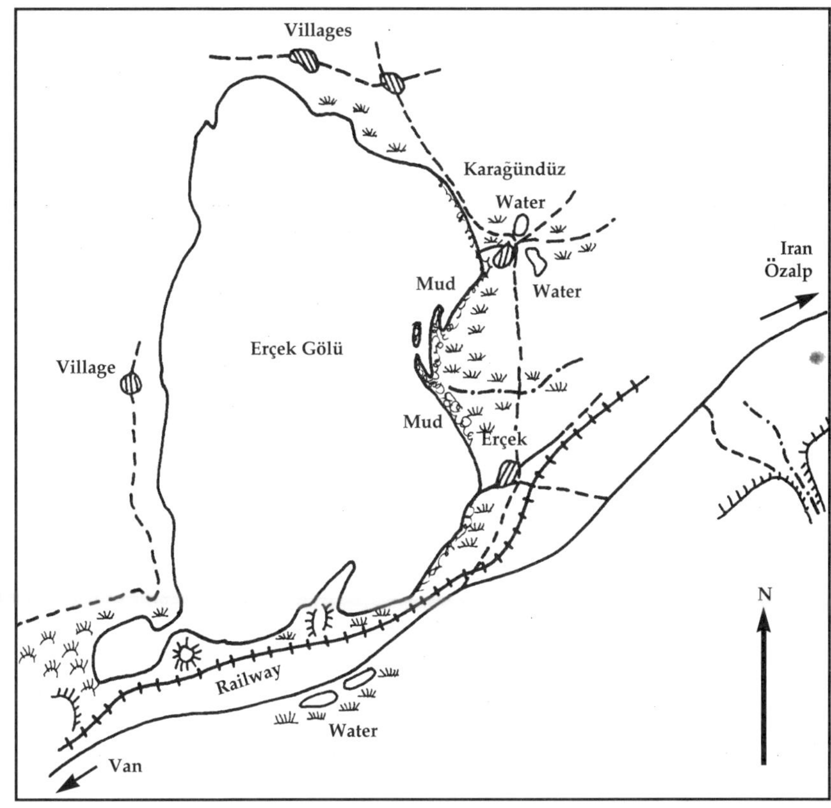

provides effluent, which, along with input from the drainage system of the valley, produces a highly productive feeding area for waders and wildfowl along the shallow eastern shore.

Location

The lake is situated about 35km to the east of Van on the road to Ozalp and Iran which skirts the southern shore. As the road starts to leave the lakeshore, there is a track leading to the village of Erçek on the left. This track continues across the valley bottom towards some barrack-like buildings on the northern side of the valley. The lake is between 500-1000m to the left all the way along this track, which then follows the more barren northern shore before continuing across the mountains to eventually meet the main road around Van Gölü.

About a kilometre to the east of Erçek village there is a small gorge on the south side of the road which can be reached along an indistinct track. Viewed from the road it appears to be a fairly insignificant gap in the hillside, but is well worth a few hours.

Dolmuşes regularly pass Erçek during daylight hours. It is possible to arrange a taxi to drop you at Erçek, and pick you up at a pre-arranged time, through the travel agent in Van (about £30), although it may be cheaper to arrange this privately.

Accommodation

There is nowhere to stay at Erçek, unless you make friends with the locals! Van is the most convenient place to stay.

Strategy

The best areas of the lake are the eastern shore and a small inlet at the south-western corner (where the lake is first sighted from the Van road).

Erçek is quite quiet during the mid-summer period, but is excellent in May, August and September. Light can be a problem in the evening on the eastern shore, this being a good time to spend a couple of hours up the gorge. In migration periods, waders and wildfowl form dense flocks anywhere along the shore from Erçek village to the northern side of the valley floor. Getting close to the birds is the only difficulty as there are no vantage points to scan from except on the main road near Erçek village. From the track across the valley floor walk, or if conditions allow, drive to the lakeshore across the grazing meadows. The lakeshore is muddy, it is unwise to walk around on it, as it very deep in places - one of the authors has a pair of shoes slowly fossilising in the muds of Erçek!

Birds

Erçek's prime attraction in spring and autumn are its waders. Avocet, Black-winged Stilt, Kentish Plover, Lapwing, Little Stint, Ruff and Redshank form the bulk, with smaller numbers of Curlew Sandpiper, Temminck's Stint, Turnstone, Marsh, Terek and Broad-billed Sandpipers, especially in autumn. In spring, Great Snipe has been recorded, and sizeable flocks of Red-necked Phalaropes have been noted. Caspian Plovers have been seen here, and may pass through the area regularly - the valley floor may be worth checking. Stone-curlew and Common Crane can be found on the grazing areas.

From June onwards Black-necked Grebe and Ruddy Shelduck numbers build up to peaks in excess of 5,000 by late summer. Smaller numbers of other wildfowl are present especially Shoveler and Garganey. Greater Flamingoes occur regularly. On the fringes of the lake Lesser Short-toed and Bimaculated Larks, Yellow Wagtail and Isabelline Wheatear are common.

The gorge and hills to the east of Erçek hold breeding Ruddy Shelduck, Egyptian Vulture and Long-legged Buzzard, with Montagu's and Pallid Harriers on migration. Eagle Owls have bred in the gorge, but may have left this site recently. Passerines in the gorge include Crag Martin, Grey Wagtail, Dipper, Black Redstart, Finsch's and Black-eared Wheatears, Rock Thrush, Rock Nuthatch, Rock Sparrow and Snow Finch. The karst-like hilltops hold Bimaculated Lark, Tawny Pipit, Grey-necked Bunting, and, with a little luck, Black-bellied Sandgrouse and Nightjar may be located.

Other wildlife

In May and early June there is a wonderful display of flowers in the damp meadows on the way to Erçek: a yellow *Pedicularis* (lousewort); a tiny rose-pink primula, *Primula algida*; and a dark red-purple marsh orchid, *Dactylorhiza vanensis?*, which has taxonomists guesing. The agamid lizard *Agama caucasica* occurs here and Jackals can be seen on an evening walk near the gorge.

Bendimahi Marsh

Bendimahi Deltası consists of a marsh complex, where the Bendimahi river enters Van Gölü. Shallow water, areas of reedbed and muddy fringes make the river delta a rich feeding ground for ducks and waders. There is a small reed-fringed inlet to the south of the river, and grazing meadows and arable land on the tongue of land in between.

Location

The marsh is about 90km north of Van at the northeastern extremity of the lake. The main road forks east for Muradiye and north to Erciş, and the first inlet is about 100m north of this junction. The bridge over the river is a kilometre to the north of this. A few hundred metres before the bridge there is a rough track towards the lake which runs through arable land almost to the shore. There are frequent dolmuşes from Van to Erciş and Muradiye, and a bus service to Erciş.

Key

Roads
River
Drivable track
Paths

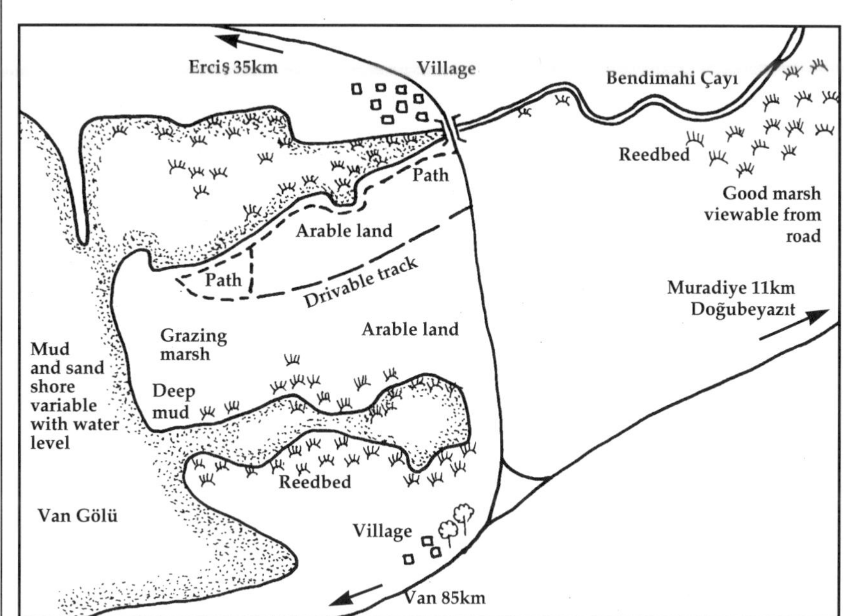

Accommodation

The road from Van is excellent and the 90km can be done in an hour, so staying in Van is quite practical. There are also some hotels in Erciş. Locally, the old garage near the bridge provides some protection from the elements. 'Şelale Waterfall' north of Muradiye is well worth staying at (see next site), where there is a restaurant, as there also is in Muradiye.

Strategy

Birdwatching is excellent from May to October. As with other sites in this region the birds present in the winter months are virtually unknown. The bridge provides a particularly good vantage point with much of the marsh visible. The reed fringes on the eastern side of the bridge are worth checking for rails. The main wader feeding grounds change according to the species present and variations in water level. One of the best areas is a five minute walk along the southern shore of the river, to a spot where the delta opens out markedly. Another good area is uon the southern side, at the point where the river actually becomes lake. Midges and mosquitoes can be a real nuisance at this site, particularly when sitting checking through wader flocks, so have some insect repellant handy. The southern inlet is difficult to work, as the reeds are tall, and usually in quite deep water. Some of the avifauna present can be seen just north of the junction, or by walking right out towards the lakeshore until the reeds thin out a little. There are several small marshes on the river upstream of the bridge which may repay further investigation.

Birds

Bendimahi is excellent for migrating waders, and along with Erçek Gölü, provides the best opportunity in eastern Anatolia to sift through the flocks, which tend to have a slightly different species composition than further west in the country. During May, Red-necked Phalaropes can be seen in quite large numbers, and while apparently scarcer in autumn, they can still be found, along with sizeable flocks of Broad-billed and Marsh Sandpipers. More common in both migration periods are Little Stint, Ruff, Green and Wood Sandpipers. Black-winged Stilt, Lapwing and Redshank are common throughout the summer.

There is a good variety of wildfowl during migration periods. Greylag Geese, Ruddy Shelduck, Shelduck, Teal, Garganey and Pochard occur in variable numbers. Ferruginous, Marbled and White-headed Ducks are present in small numbers, most commonly in the southern inlet - all three probably breed here.

Summer, and early autumn records of boreal breeding species such as Black-throated Diver, Goldeneye, Goosander, and Pomarine and Arctic Skuas are surprising but not unprecedented in the Van basin.

Virtually all the Turkish heron species have been recorded here, although usually as singles, or small groups. Pygmy Cormorants occur throughout the non-winter months. White Pelicans, in flocks up to a thousand, and Greater Flamingoes are irregular but frequent visitors, presumably from the breeding colonies at Daryachek-ye-Reza'Iyeh (Lake Urmia) in Iran. Common Cranes are also often present and probably breed nearby.

Eight species of tern have been observed around the estuary, the marsh terns being the commonest, especially on migration. Caspian Terns can be found along with *'armenicus'* Herring Gulls and smaller numbers of Slender-billed and Mediterranean Gulls roosting on the sandy spits and islands at the rivermouth.

The arable fields contain lots of Calandra and Lesser Short-toed Larks, and Corn Buntings with Yellow Wagtails breeding on the grazing marshes. Very large numbers of these pass through in autumn, mostly *M.f. beema* and *M.f. feldegg*. Citrine Wagtails probably breed, and should be looked for in the damper fields, a habitat also favoured by Red-throated Pipits on migration.

More unusual records include Great Bustard, Black-winged Pratincole, Great Snipe and Terek Sandpiper, the last three probably occuring in small numbers annually.

Other wildlife

Both Stripe-necked and European Pond Terrapins occur and there is a good show of orchids in spring.

Şelale Waterfall (Gönderme) and Çaldıran Ovası

As well as being an enjoyable place to stay, and handy for Bendimahi, Şelale Waterfall provides an opportunity to see many of the characteristic species of mountain and high-steppe habitats in eastern Anatolia. In particular it gives birdwatchers the chance to get to grips with two difficult species - Saker and Eagle Owl. The marshes of Çaldıran Ovası and the jumbled black lava slopes of Tendürek Dağı to the north provide scope for discovery, but are in a sensitive area alongside the Iranian border.

Location

Şelale Waterfall is situated about 6km north of Muradiye on the road to Çaldıran and Doğubeyazıt. Here the river Bendimahi plunges 20m into a small gorge, resulting in a quite spectacular waterfall. There is a rather worrying rope bridge swaying above the gorge which leads to the restaurant. Just north of Çaldıran the Bendimahi valley opens up to a flat plain with a series of damp areas, marshland and meadows. This is Çaldıran Ovası. The area is best explored by car, since bus services are infrequent. Recently though, the road past Çaldıran has been improved and this may result in better public transport. It is not too difficult to get to the campsite from Muradiye or Van, although you may have to wait for a bus. Many tourist buses stop at Şelale Waterfall and it is possible to hitch from there.

Accommodation

At the time of writing it is possible to camp or to stay at some small chalets available at Şelale Waterfall. It is usually possible to sleep in either the restaurant or in the Carpet Co-operative building, especially after a salubrious evening with the locals (and the occasional truck driver) in the restaurant. There is a hotel under construction next to the restaurant. Currently, the place is just a meal stop-over for tourist coaches and truck drivers on the Van to Doğubeyazıt route. The restaurant is good and inexpensive, although power cuts and food shortages are not unknown. There are plenty of fish in the river, and these frequently find their way onto the menu.

Alternatively, the area can be worked from Doğubeyazıt or even Van, depending on the length of time available.

Strategy

May to June, and mid-August through until October are the best birdwatching periods, although birds such as Saker and Eagle Owl can be found at any time.

The Bendimahi valley is a minor raptor flyway. Sitting outside the restaurant with plentiful supplies of tea and a watchful eye on the sky

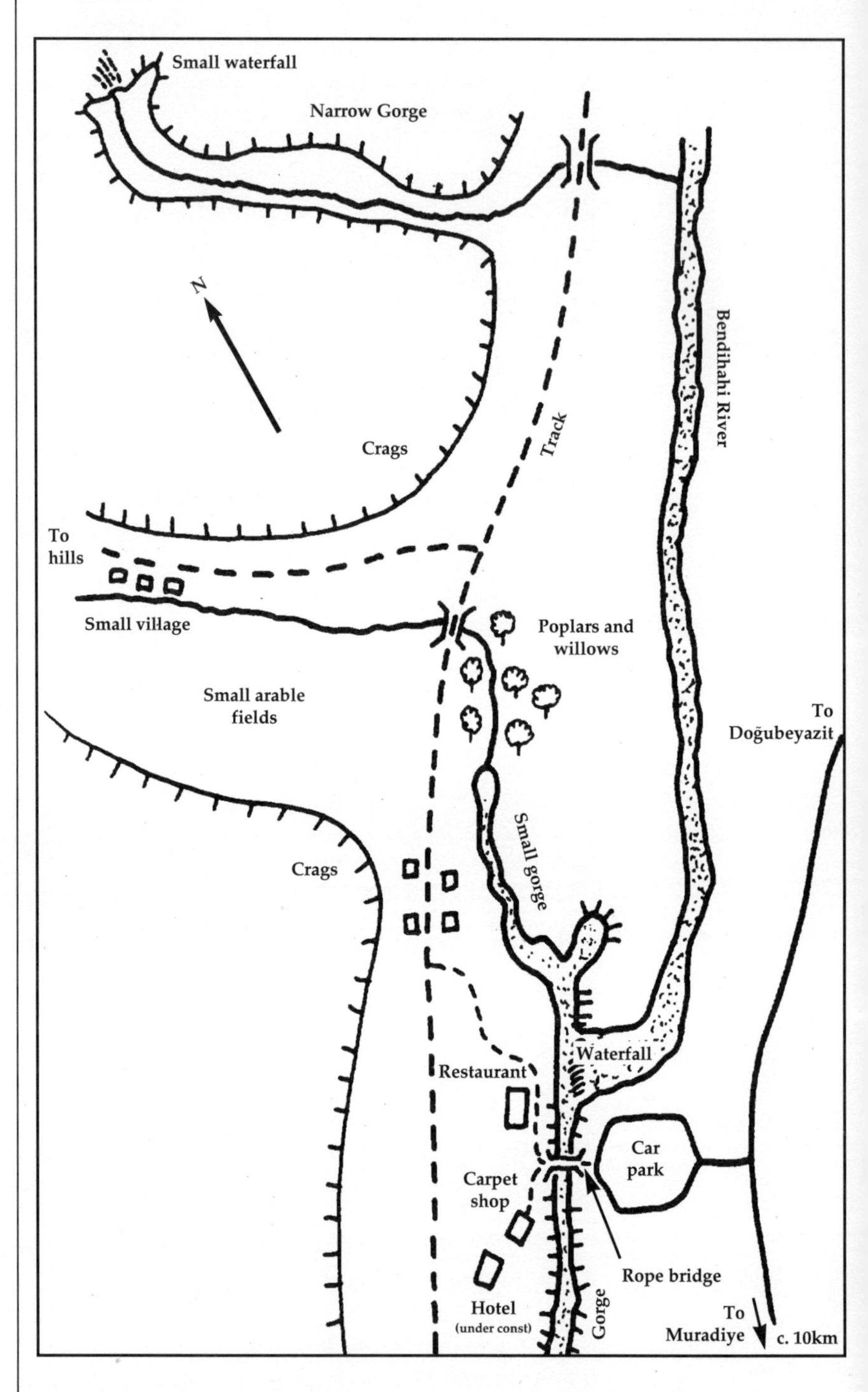

should produce a number of species in early spring or autumn. The poplars and willows just above the waterfall act as a migrant trap, and on good days can have plenty of interest, particularly various *Phylloscopus* warblers. Shortly after dawn is best as the birds seem to start moving on again by mid-morning.

The valley to the west, behind the restaurant and poplars, has some cover around a small village and there are some good areas at the head

White-throated Robin

of the valley. The plateau is quite barren once the greenness of spring has passed, and the avifauna consists largely of larks and wheatears. The second valley to the north of this is more gorge-like and rises rapidly through a series of (often dry) waterfalls onto the plateau. An evening visit to this valley may produce Eagle Owls.

The marshes near Çaldıran can be checked out from the road or any of the numerous tracks into the meadows, but extreme caution must be exercised in the area, because it is right alongside the border with Iran. Checks on vehicles are regular when leaving Çaldıran (or on the road outside Doğubeyazıt) and if you are stopped and intend to watch along this road, it would be wise to tell the soldiers that you will be doing so in case you are stopped later. The road is also supposedly closed at night. Having said this, the authors have had no problems in the area.

Birds

Sitting outside the restaurant watching for raptors should produce local breeding species such as Egyptian Vulture, Long-legged Buzzard, and Hobby. With a little luck, Saker may be seen hunting low along the crest of the plateau behind the restaurant. In autumn species seen on passage may include; Montagu's and Pallid Harriers, Sparrowhawk, Buzzard, Golden Eagle (may well be resident), Osprey and Peregrine. These were recorded in a couple of days by the authors and probably don't represent a genuine cross-section of raptor migration through the valley.

The wooded area above the waterfall holds Hobby, Turtle Dove, Cuckoo, Hoopoe, Golden Oriole and Cetti's Warbler, while migrants noted here include Wryneck, Spotted and Semi-collared Flycatchers, Savi's, Marsh, Olivaceous, Upcher's and Barred Warblers, Whitethroat, Lesser Whitethroat, Garden Warbler, Blackcap, Wood Warbler, Mountain Chiffchaff, Chiffchaff, (many of which appear to be of the *'abietinus'* group) and Willow Warbler.

The fields, trees and steep hillsides up the first valley are a good place for seed-eating passerines. Rock Sparrow, Goldfinch, Linnet, Common Rosefinch and Black-headed Bunting are all quite easily found. As the valley narrows and becomes rockier, Black-eared and Finsch's Wheatears, Rock Thrush, Black Redstart (an eastern race) and Rock Nuthatch become the predominant species. The particularly striking *'variagata'* race of Stonechat can be found on the steep slopes below the plateau, and on the plateau itself the commonest species are the Bimaculated Lark, Tawny Pipit, wheatears and Crimson-winged Finch. During migration plenty of Whinchats, Northern Wheatears and Ortolan Buntings stop to feed on the slopes.

The second valley has similar species, with Dipper, White-throated Robin and Rock Bunting being easier to find along the stream course, and Long-legged Buzzards nesting on the low cliffs at the side of the valley. Eagle Owls are best looked for around dusk, especially at the open end of the gorge, although they can be seen anywhere in the vicinity - even perched on rocks in the river. It is imperative that local hunters are not informed of the presence of Eagle Owls anywhere in the country as there is (for locals) a very lucrative trade in selling these birds to Arab countries. They seem to disappear from sites known to birdwatchers, and recently captured birds have been seen in Van. Other species noted in the general vicinity include Woodcock, Common Sandpiper, Alpine Swift, Crag Martin, Red-throated Pipit, Citrine and Grey Wagtails and Desert Warbler (vagrant).

The marshes are good for Common Cranes and various wetland species can also be found, including Pygmy Cormorant and, unusually for this region, Bittern. The jumbled lava slopes of Tendürek Dağı to the north of Çaldıran hold good numbers of Snow Finch, Twite and Crimson-winged Finch. Mongolian Trumpeter Finch has been reported recently.

Other wildlife

Jackal and Wolf occur in the mountains surrounding this area but are difficult to see. The stunning bright red parasitic plant, *Phelypaea coccinea*, a relative of the broomrapes, can be seen flowering above the waterfall during May. Also growing here is a rare and beautiful iris of the *'paradisea'* group, with large white and velvety purple-brown flowers.

Doğubeyazıt

Many visitors to Turkey go to Doğubeyazıt to see the wonderful panorama of Ararat and to visit one of the country's premier historic

Grey-necked Bunting

sites the Işak Paşa Palace, but, as far as birdwatchers are concerned the location was of no particular importance until the discovery of a population of Mongolian Trumpeter Finches there during 1992.

Location

The Işak Paşa is 6 kilometres east of Doğubeyazıt on the lower slopes of some very craggy hills. It is well signposted from the town with a tarmac road leading to the palace itself. The site is well visited by tourists and as a result there is a tea house/restaurant of distinctly western looking design situated directly above the palace. There is a small fee if you should desire to look around the palace, which is quite stunning, as much for its setting as for the well preserved buildings.

Accommodation

Doğubeyazıt is on the major trade route through to Iran and eastwards, and contains a large number of hotels, some of which cater for tourists requiring a degree of comfort. The best (and most expensive) is the Hotel Isfahan, located just off the main street near the bus station. There are also a number of cheap but quite good establishments on the main street, the best of which are set back, up side-streets, with signposts indicating their locations. Restaurants and kebap houses are plentiful; the best restaurant in town is in the Hotel Isfahan, but this closed (permanently?) in late 1992. There is a campsite up by the Işak Paşa.

Strategy

The palace is backed by hills, into which a track can be followed to some ruins (visible from the palace), which are built into the crags. Beyond this point the track winds up along the base of some scree slopes composed of small greenish-coloured rocks. The best areas to look for the Mongolian Trumpeter Finches are on these sparsely vegetated slopes and along the base of the cliffs above. They are relatively tame and approachable. The whole area would benefit from further exploration, but if you do decide to wander further afield here remember that you are very close to the border with Iran!

Birds

As already mentioned this is the only currently known regular site for Mongolian Trumpeter Finch in Turkey, although there have been one or two other records recently. So far they have been recorded here between June and late August, and although juveniles were present in August, it is too early to be sure whether they breed here or not. However, they have been spreading across Iran and it seems quite possible that the Turkish observations are the start of colonisation. Other species of interest include another eastern speciality, Grey-necked Bunting, which can be seen on the same slopes, along with Crag Martin, various wheatears, Blue Rock Thrush and large numbers of Chough on the crags themselves.

The Aras valley, on the border with Azerbaijan and to the northeast of Ararat, is a little known area ornithologically. Blue-cheeked Bee-eaters probably breed in the valley, and Pale Rock Sparrows have occurred on the severely eroded, bare hillsides near Tuzluca.

Arın Gölü and Suphan Dağı

Arın Gölü (also known as Sodalı Gölü) is a small soda lake separated from Van Gölü by a narrow stretch of land. The mineral content of this lake is much higher than the main lake and as a result there is virtually no emergent vegetation, except for the small marsh and the wet grazing meadows by the village of Gölduzu. Behind Arın, on the other side of the main road, the massive bulk of Turkey's second highest volcano rises to 4434m (although some maps put it at 4058m). Suphan Dağı is largely barren with some vegetation cover in the bottom of gullies.

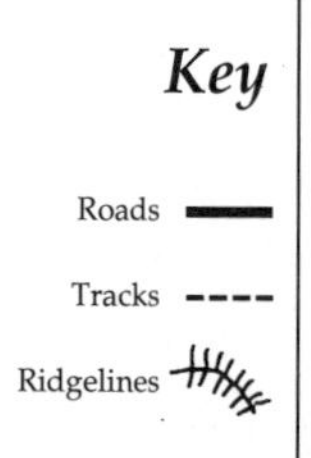

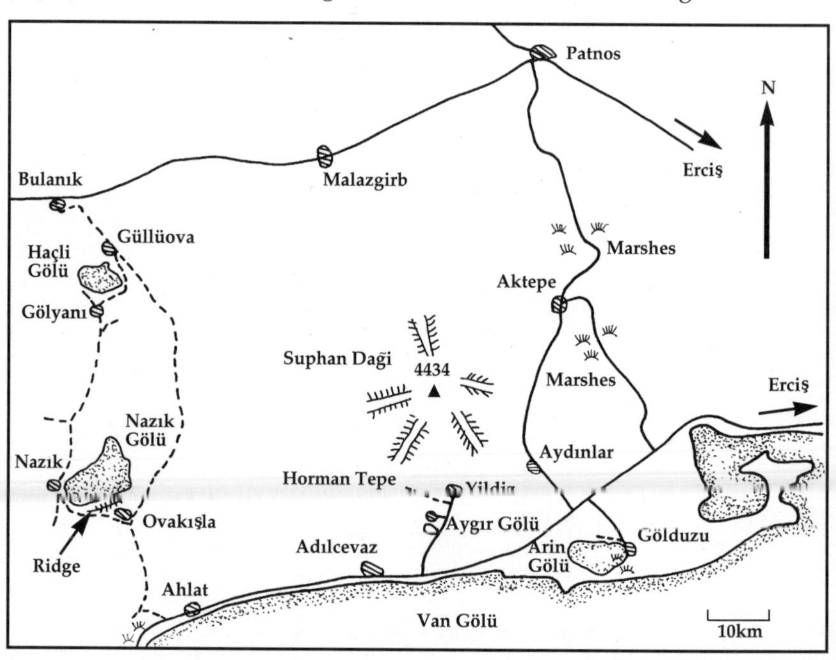

Location

Arın Gölü is situated south of route 965, about 30km to the east of Adılcevaz, on the north side of Van Gölü. Take the southbound road (partially tarmac-covered) signposted to the village of Gölduzu. The best areas are within a few hundred metres of the village. The mountain is best reached from Adılcevaz (see strategy). There are dolmuşes and buses to Adılcevaz from Erciş and Tatvan and these will also drop you at the Gölduzu junction. The best way to get to the bottom of the mountain is by taxi from Adılcevaz and a return pick up can also be arranged.

Accommodation

Adılcevaz has the nearest hotel accommodation and provides a good base for Suphan Dağı. The Hacıbaba Turistik Oteli, at the east end of the main street is used by climbers and walkers and is of great help in arranging transport to the trailhead on Suphan Dağı, and for information on the mountain. Alternative accommodation can be found in Eráis. Adılcevaz has a few restaurants, all are adequate but fairly basic and the shops are of similar quality. Ahlat, some 20km west of Adılcevaz, has the rather imposing looking Selcuk Hotel standing on its own lakeshore by the main road. This is more comfortable than most hotels in the region and is a good central base if you plan to visit Arın Gölü, Bulanık, Haçlı, Nazık and Nemrut Dağı.

Strategy

Arın is at its best during May and June, and from August to early October, as it is primarily a stopover point for migrating waders and wildfowl. Any time of day is reasonable, except in late evenings when the sun will be directly in front when looking out onto the water.

The marshy area is easily checked from the village, although children will follow you everywhere (and some adults). The best, and most peaceful way to work both the marsh and lake, is to take a track to the right just as the first houses are reached on entering the village. This track peters out onto a flat grassy area at the northern end of the marsh. From here a walk down the salt encrusted sand/mud bar, will take you to a small hillock opposite the main village. This is an excellent all round vantage point, and although you may have to share it with cattle, the number of accompanying children should be reduced.

For the long, but walkable trek up Suphan Dağı, arrange a taxi from the hotel in Adılcevaz. It will take you to Hormantepe village before dawn, or alternatively start equally early from Aydınlar. If you intend to reach the top, it is a long hard slog, and it is best to hire a local guide, although this is not absolutely necessary. There is an excellent description of Suphan Dağı and details of the best routes up the mountain in 'Trekking in Turkey' by M. Dubin and E. Lucas (Lonely Planet Guides, 1989). It is worth getting to the summit for the breathtaking views over Van Gölü. However, the best birdwatching is to be found on the first two-thirds of the ascent. Passerines are best looked for among the sparsely vegetated gullies. In autumn, there seems to be a regular south-westward drift of migrating raptors, across the south facing shoulders of the volcano.

Birds

In late summer and autumn Arın Gölü is host to large concentrations of wildfowl. Thousands of Black-necked Grebes, Pochards and Coots cover the lake, and this is clearly an important stopover point for White-headed Duck, with up to a thousand being present during August and September. Several other species are usually present in small numbers including Ruddy Shelduck and Ferruginous Duck.

In summer, large numbers of terns and gulls can be seen, often sitting on the sandy bar at the entrance to the marsh. Black-headed, Herring (*armenicus*) and Mediterranean Gulls, and Caspian, Common, Little, Whiskered and White-winged Black Terns forming the bulk. The marshy area is excellent for waders. Black-winged Stilt, Lapwing, and Redshank breed, while in autumn hundreds of Little Stints, and smaller numbers of Broad-billed Sandpiper, Ruff, Snipe, Black-tailed Godwit and Marsh and Wood Sandpipers are usually present. The marsh and lake fringes are rather poor for passerines, although Yellow Wagtail is common, and Hoopoe, Short-toed and Lesser Short-toed Larks and Isabelline Wheatear are found around the village. As with other Turkish wetlands in late summer, phenomenal numbers of Sand Martin may be present, the sky can often be partially obscured by their sheer density. On one visit to Arın an estimated 20,000 were seen over the 40 hectares of marsh.

Suphan Dağı is little known, ornithologically speaking. The mountain, particularly between 2,000 and 3,000m holds good populations of Shore Lark, Tawny Pipit, Black Redstart, Isabelline and Finsch's Wheatears, Rock Thrush, and Spanish Sparrow. Radde's Accentors are quite common, and during the autumn flocks can be seen going to roost in areas of damp vegetation in the gullies above Hormantepe. Crimson-winged Finches should also occur here. The only birds noted on the exposed summit area (over 4000m) were Lesser Kestrel and Snow Finch. Autumn migrants using the mountain slopes include Nightjar, Red-throated and Tree Pipits, and large numbers of Yellow Wagtails (mostly of the *beema* race). Migrating raptors include good numbers of Buzzard (*B.b. vulpinus*), various *Aquila* eagles and harriers. Lammergeier, Long-legged Buzzard, Golden Eagle, Saker and Peregrine all occur here, and probably all breed locally.

Nemrut Dağı

The eruption of Nemrut Dağı caused the formation of Van Gölü when lava flows and volcanic debris blocked the natural drainage outflow of the basin. The eruption must have been extremely spectacular judging by the remaining caldera on Nemrut Dağı. The sheer volume of missing mountain is apparent when standing on the rim of the ten kilometre-wide crater. Cliffs up to a thousand feet high plunge into the main crater lake on the north side of the volcano, while vast blocks of black obsidian, and cuttings in the layers of volcanic dust testify to the forces which were unleashed here.

Location

The crater is situated north of Tatvan on the western edge of Van Gölü. Take the Ahlat/Adılcevaz road from Tatvan, and a few kilometres along here there is a track signposted to Nemrut Dağı (yellow tourist sign), next to a Petrol Ofisi garage. From here it is about a half hour drive to the crater rim. The tracks are perfectly drivable, but be careful on areas of volcanic ash, as often half the road has disappeared into a void. You can drive into the crater, where a fork to the left will take you down the edge of an obsidian-littered valley towards the main crater lake, and a right fork, which goes past various tented encampments to the hot springs and eventually to the crater lake again.

Tatvan and Bitlis can be reached by coach services from most major towns, and the former by rail from the west and by ferry from Van. Dolmuşes, which ply the crater tourist trade, can be taken from Tatvan.

Accommodation

Camping out in the crater is possible, but it does get quite cold at night. The nearest place to stay is Tatvan, where there are a number of hotels and restaurants on the main street, most of which are adequate. About 20km to the southwest is Bitlis, a town seemingly perched on the edge of an abyss. Here the Hanedan Otel on the main `through' street is quite good for around £4 a night. This hotel was opened in 1988 and even has western style toilets! The hotel restaurant has good quality food and a wide range of dishes, and it is reasonably cheap.

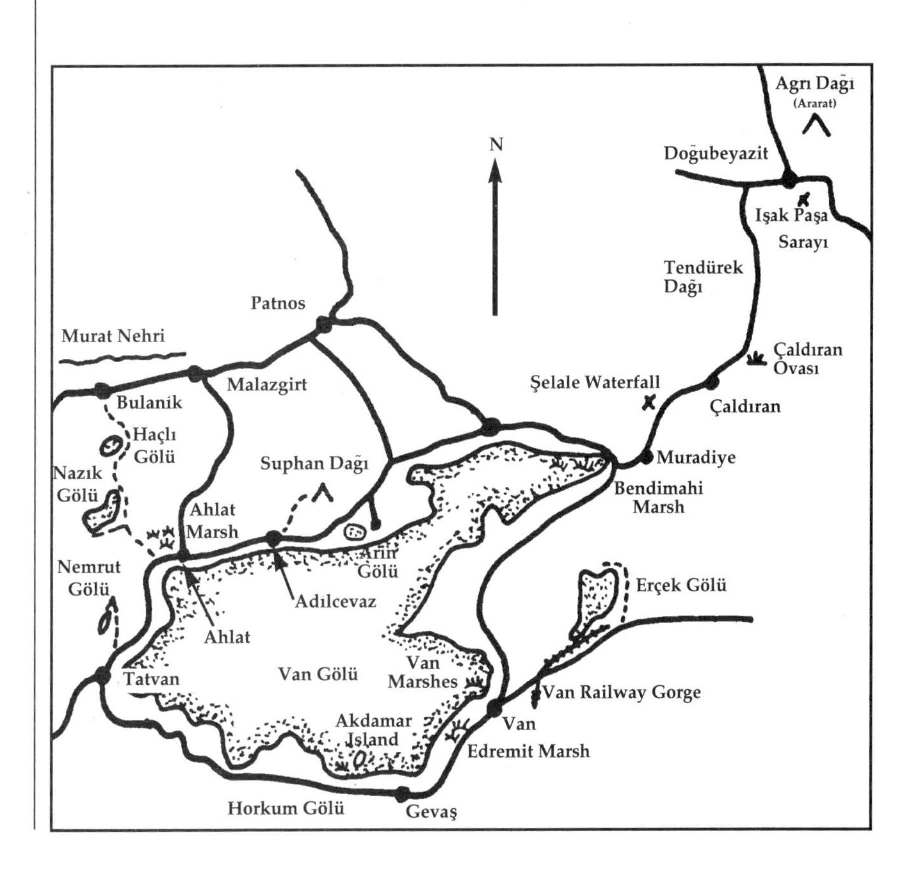

Strategy

For most of the species present, a wander around any part of the crater should be sufficient, the most productive areas being along the ephemeral watercourses and the sparsely vegetated slopes of the crater.

Birds

Nemrut Dağı produced one of the more unusual discoveries of Turkish ornithology - a breeding population of Velvet Scoter. They have since been found at at least two more high altitude lakes in north-eastern Turkey. There are usually at least half-a-dozen pairs present in the summer months. Common breeding species on the crater include; Bimaculated Lark, Shore Lark, Crag Martin, Tawny Pipit, Black Redstart, Isabelline, Black-eared and Finsch's Wheatears, Rock Thrush, Mistle Thrush, Rock Nuthatch and Rock Sparrow. Radde's Accentors breed in small numbers - one of the best areas to look is among the jumble of obsidian boulders. Chukar are quite common and are best located by call around dusk or dawn, especially in the low scrub bordering the lake. Grey Partridge, Red-tailed Wheatear, Bluethroat, White-throated Robin, Red-fronted Serin and Grey-necked Bunting have all been seen in summer, and may breed in small numbers.

Bee-eater, Yellow Wagtail and Red-throated Pipit are likely to be encountered on migration. In autumn, there is a focusing of raptor migration along the northern shore of Van Gölü, funnelling through the Tatvan area. This can be seen across the south-eastern shoulder of the mountain. Locally, Long-legged Buzzard and Egyptian Vulture breed, and Griffon Vulture, Golden Eagle and Eagle Owl probably do so. More unusual records here include White-tailed Plover, Egyptian Nightjar, Citrine Wagtail, Ring Ouzel, Trumpeter Finch and Cinereous Bunting.

Other wildlife

The surrounding slopes, especially to the west of the mountain, have a very distinctive flora dominated by several species of shrubby oaks. Growing amongst them are a variety of orchids *(Orchis umbrosa, Orchis caucasica, Orchis pinetorum* and *Epipactis latifolia)* - irises and paeonies. The crater itself has an endemic flax and an endemic buttercup! Two species of whip snakes, *Coluber najadum* and *C. ravergieri*, can be found inside the crater as well as Snake-eyed Lizard *Ophisops elegans, Lacerta media*, and the small frog *Rana camerani.*

South Van Gölü

Between Van and the village of Göründü, there are a number of small wetlands on the edges of the main lake. These are of interest primarily for their dense breeding populations of various wetland species, in particular Ferruginous and White-headed Ducks.

The best lagoons are both by the main road around Van Gölü, one is near Edremit about 10km south of Van, and the other is about 20km west of Gevaş near the village of Göründü. Both are only a few hectares in extent. Other places worth a look are the rivermouth at Gevaş, and the mouth of the Güzelsu, northwest of Gevaş. These areas are easily visited from Van.

There are two campsites, Güney Kamping about 25km south of Van, and Cafer Camping west of Gevaş. Both are cheap, have a restaurant on site, but are not always open. In spring or autumn the trees on both campsites are excellent migrant traps and the following were recorded on two autumn days at Cafer Camping and the quarry in the hillside above; Nightjar, Wryneck, Thrush Nightingale, Marsh, Reed, Great Reed, Olivaceous, Bonelli's and Wood Warblers, Golden Oriole, Cinereous Bunting and forty Grey-necked Buntings.

Both lagoons are easily visible from the road, although to have a good look at the one near Edremit, it is necessary to walk up the sandbar seperating the lagoon from the lake. Breeding species include Great Crested, Red-necked and Little Grebes, Little Bittern, Ruddy Shelduck, Pochard, Gadwall, Ferruginous and White-headed Ducks, Black-winged Stilt, Kentish Plover and Great Reed and Moustached Warblers. Black-necked Grebe, Pygmy Cormorant, Night Heron, White Stork, Marsh Harrier, Hobby, and various waders and marsh terns are usually present. Summering records of boreal wildfowl such as Goldeneye, Long-tailed Duck and Velvet Scoter have been noted. These individuals often appear very pale and washed out.

Akdamar Island (also spelt; Ahdamar, Ahtamar or Achtamar).

This is a small island to the northwest of Gevaş. There is the tenth century church of the Holy Cross to lure tourists, and as a consequence, there are regular boats going to the island from opposite the motel to the west of Gevaş. From a birdwatcher's point of view, the island holds colonies of Chough and the '*armenicus*' race of Herring Gull on the cliffs at the west end of the island. There is also a small colony of Night Herons in a couple of trees at the base of the cliffs. There is some good cover for migrants on the island and Alpine Swift, Red-backed Shrike, Golden Oriole, and *Phylloscopus* warblers are common in autumn.

Bulanık

The Murat Nehri valley, between Bulanık and Balato, is one of the few remaining breeding sites for Demoiselle Crane in Turkey. The wide floodplain of the Murat is farmed for various crops, especially melons and sunflowers, and damp hay meadows, most of which are still harvested by hand provides an opportunity to look at a rare and declining habitat. The higher ground to the south of the river is criss-crossed by irrigation channels and is almost wholly given over to wheat. Haçlı and Nazık Gölüs are two reaonably large lakes between Bulanık and Van Gölü. Haçlı is shallow with muddy fringes and grazing marsh surrounds which are full of birds, especially during migration. Nazık is deeper with very little vegetation, and is little known.

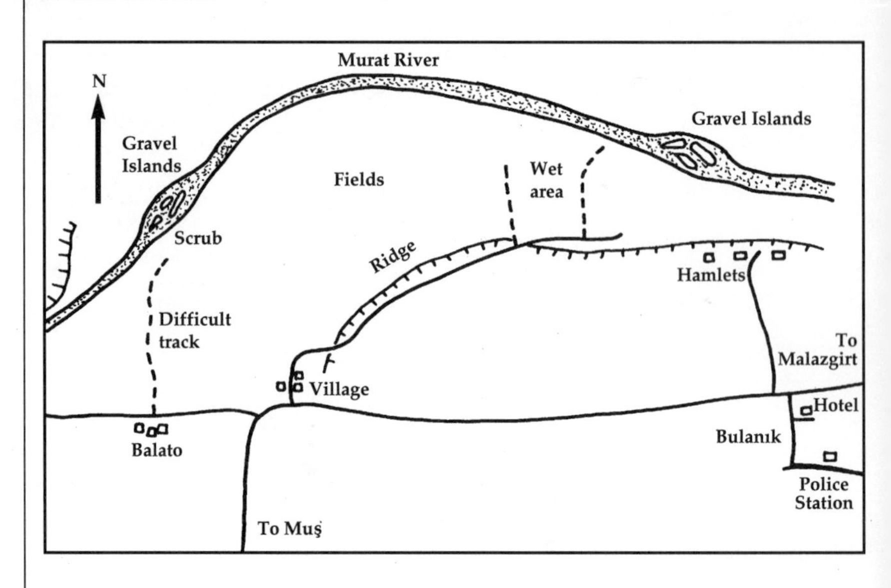

Location

Bulanık is about 60km west of Patnos on route 280. To get to the floodplain it is necessary to cross the rolling arable land around Bulanık. The easiest access is 10km west of Bulanık, on the Varto road. Just past the junction where the main road curves south towards Muş, and opposite the village of Balato, there is a track which goes across the floodplain, and depending on how dry it is, almost reaches the banks of the river. In Bulanık there is a track (next to a large orange building), which eventually tops the scarp overlooking the river. Various tracks go along the edge of the scarp, and down to the riverbank from here.

The only access to Haçlı and Nazık is by dirt tracks, which are driveable for most of the summer. To reach them from the south, follow the track to Ovakı şla (currently being improved), which leaves the Van Gölü 'ringroad' about 2km west of Ahlat. Head left at Ovakı şla and this track eventually skirts the western edge of Nazık Gölü, leading to the village of Nazık. To continue on to Haçlı Gölü, follow the track due north and turn right at the only obvious junction at the head of the valley containing the lake. To find Haçlı from Bulanık involves locating the correct track out of town. Head southeast out of town past the police station for about a kilometre, keeping left at a fork in the road. After passing a small pond on the left, take the next right track southwards along the valley, keeping right at the first proper village (Güllüova).

Bulanık is a bit off the beaten track and therefore not easy to get to by public transport. Dolmuşes between Muş and Patnos pass through and are the easiest towns to reach Bulanık from. The only way to get to Haçlı and Nazık without a car is by infrequent dolmuş and hitching.

Accommodation

The only town with hotels in the area is Bulanık, and they are very basic. There are at least two in the centre of town, and several restaurants, which are also rather basic. Electricity and water supplies are often cut in eastern Turkey, particularly around Bulanık.

Demoiselle Cranes

Strategy

The prime aim of most birdwatchers visiting Bulanık is to locate the Demoiselle Cranes, and this is not always easy. They are present from mid-May to the end of August. There are usually one or two pairs, although exceptionally 22 individuals have been recorded. The Demoiselles (and also Common Cranes) often frequent, and probably breed on, the gravelly islands in the river. Thus, the best areas to look are the group of islands near the village on the edge of the scarp at the Bulanık end and the island at the Balato end. If they aren't located in either of these areas then they may be feeding anywhere on the floodplain, which can be checked on foot (very hard work) or by driving along the track on the edge of the higher ground, from the village of Yoncalı to the village at the Bulanık end. The river forms a large arc here, and the central portion is wide and difficult to check, however, regular scanning should eventually locate these beautiful birds. Most of the rest of the species occur over the whole area, except the terns which breed on the islands at the Bulanık end. Any damp areas will contain waders depending on the season.

Haçlı is very open and easily worked, as is Nazık. Nazık's shoreline is mainly rocky and the best areas are at the western end where there is shallow water and emergent vegetation.

Birds

As well as Demoiselle Cranes, there are up to 10 pairs of Common Cranes breeding in this section of the valley, and considerable numbers of the latter congregate in the area from July onwards. Other wetland species usually present include; Pygmy Cormorant, Glossy Ibis, Greylag Goose, Ruddy Shelduck, and small numbers of various heron and duck species. The damp fields hold Oystercatcher, Black-winged Stilt, Lapwing, Redshank, and small numbers of migrant waders.

The gravel islands have breeding populations of Gull-billed Tern and perhaps Caspian Tern and Stone-curlew. There are plenty of harriers about, Marsh Harriers on the floodplain and Montagu's Harriers quartering the arable land, where Quail are also common. Long-legged Buzzards and Hobby can often be seen hunting over the floodplain.

Calandra and Lesser Short-toed Larks and Stonechats are reasonably common, as are Marsh Warblers in the limited areas of scrub along the river. Rose-coloured Starlings have been noted more regularly here than elsewhere in Turkey, but their appearance is very unpredictable, some birdwatchers recording large numbers, with others, even in the same year, seeing none.

Great Bustard, Black-winged Pratincole, and Black-bellied Sandgrouse may breed locally, while uncommon passage species include Great Snipe and Blue-cheeked Bee-eater.

Nazık has in the past had good numbers of White-headed Duck, however there have been no recent records. Dalmatian Pelican has been recorded, and a recent visit yielded Pygmy Cormorant, Ruddy Shelduck and more than 1000 Garganey. The surrounding steppe holds Great Bustard and Saker.

White Pelican is regularly recorded on Haçlı and Dalmatian Pelican has been seen occasionally. In late summer and autumn there are flocks of herons present, especially Little Egret and Spoonbill, along with Pygmy Cormorant, Greylag Goose and Ruddy Shelduck. Common Crane and Marsh and Montagu's Harriers are often present, and many migrant waders use the lake. Black-bellied Sandgrouse are common in the area, and can often be seen flying in to drink. Steppe Eagle, Pallid Harrier and Saker have all been recorded in the area.

Ardahan Ovası and Çıldır Gölü

Situated between the Yalnızçam Dağları and the Armenian border is a large expanse of montane steppe, generally between 2,000 and 3,500m in altitude. The landscape is a mixture of undulating steppe-covered hills, cut occasionally by spectacular rocky gorges, and wide flat valley floors. The latter contain an interesting mosaic of hand-cut hay-meadows and damp marshland which harbour a good mixture of birds. The hills to the east and south of Ardahan have large areas of rather open Scots Pine forests which hold a substantial raptor population. Çıldır Gölü is a large lake situated right on the border with Armenia.

Location

This area can be approached by road from the Artvin/ Şavşat direction in the west. This takes you over the Çam Geçidi, a 3,000m pass which can be treacherous even in the height of summer. It is easiest to approach from Kars to the south. This is a good area to birdwatch by literally just wandering around. Several areas of wetland are well worth a look. The road westwards out of Ardahan takes you across a flat valley floor which is very productive, especially where the road crosses a stream by the village of Sulakyurt at the foot of the mountains. The valley of the Kura Nehri to the south of Ardahan contains a lot of good habitat and can be seen from the Göle road near the town itself and further south by taking a side road to Yalnızçam, crossing the river, and then following a rough track northwards along the west side of the river. At Yalnızçam, the road appears to continue straight on across the river, but in fact, at the time of writing it stops at

Key

Rivers
Roads, tarmac
Drivable tracks

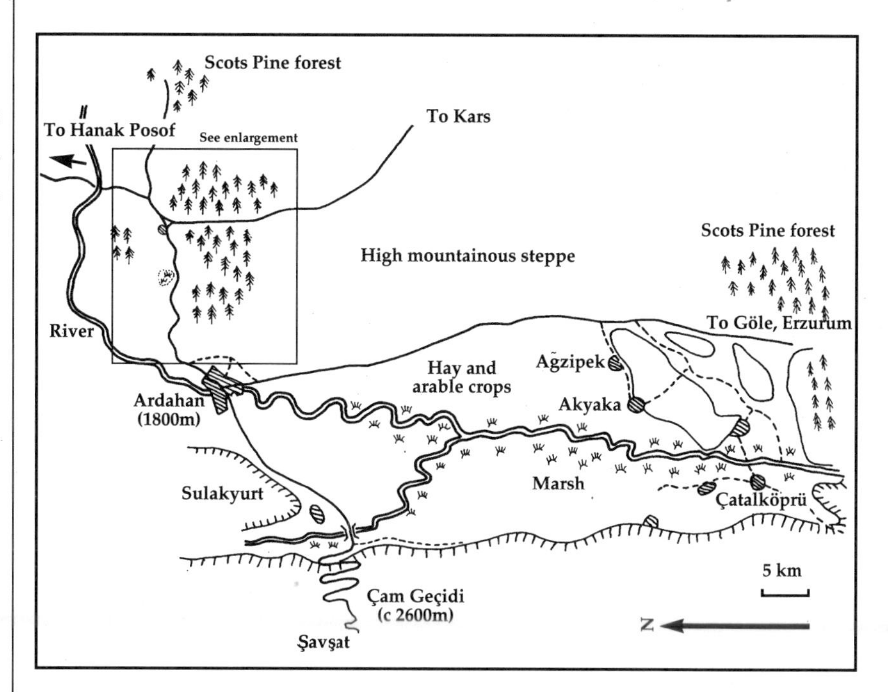

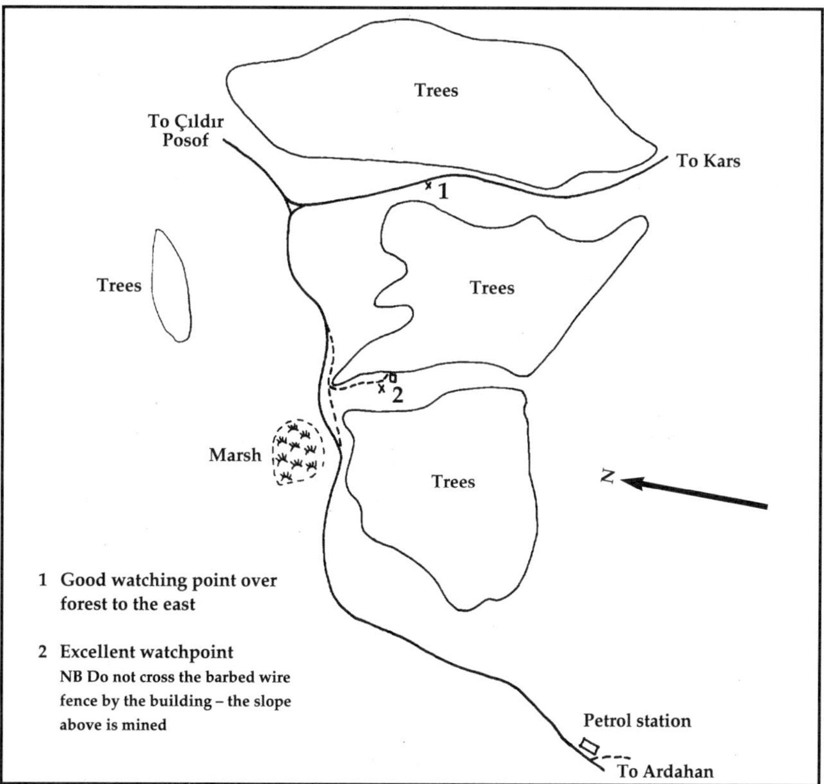

the river, so take the dirt track towards the obvious village on the far side as there is a bridge on this road.

One of the best areas can be reached by taking the Çıldır/Hanak road northeast out of Ardahan. After about five kilometres there is an excellent small marshy area in a depression to the north, right next to the road, and stretching up to a ridge to the south is some Scots Pine forest. Raptors can be seen above this forest, all the way along to the junction with the Kars road, and also for about three kilometres south along the Kars road.

Çıldır Gölü is best approached by following the road which runs round the northern and eastern sides of the lake and stopping at good spots along the shore. There is a rugged gorge a few kilometres to the west of the town of Çıldır on the Ardahan road. A 'fairy-tale' fortress can be seen on a rock pinnacle jutting out into the gorge and must be one of the most spectacular ancient sites in Turkey. Take great care in this region - we found new rocket launchers in the bottom of this gorge!

Buses and dolmuşes travel on most of the roads in this area, but they are not all that frequent. Buses go to Ardahan from Şavşat, Artvin and Kars. The roads are rather quiet, but hitching is still likely to be a good option in this area, although the political situation in the region should be carefully considered before deciding on this method.

Accommodation

Ardahan has a couple of rather poor quality hotels, the biggest of which is, when coming from the west, past the castle, over the bridge and two hundred metres directly up the hill. Otherwise the nearest accommodation can be found in Kars, which has a number of hotels of varying standard, mostly rather dirty and dingy, or in Şavşat, which has several hotels. These are generally clean and simple, but have gone downhill fast in recent years. There has been a large influx of Russian prostitutes who have taken over all of these hotels except the Hotel Kent - so far! There are plenty of restaurants in Ardahan, and a lakeshore 'cafe' just off the road at the southern end of Çıldır Gölü.

Strategy

This is a very cold region during the winter months and is usually covered in snow. The best time to visit is between late May and October. The sizeable population of breeding raptors is augmented in spring and autumn by large numbers of passage migrants, which tend to concentrate along the ridgelines in daytime. In the evenings and early mornings many of these birds descend into the valleys to roost, indeed in autumn it is not uncommon for almost every hayrick to have a raptor sitting on it. The afternoon is a good time to watch for the raptors which breed in the forests, as they soar over the tree tops. There are two particularly good spots to watch for raptors. The first is from the Kars road south of the junction with the Ardahan/Çıldır/Hanak road. Anywhere along the first few kilometres of this section is good. The second is found by taking the dirt track next to, and on the south side of, the main road by the small marsh a few kilometres east of Ardahan. This runs along the edge of the forest before reaching a clear area with an obvious derelict building on the slope above you. The track runs up to this building and you can park here. Watch from the open area below the building - don't stray across the fence or above the building, as there are still live mines in that zone! This is a superb spot to get really close views of the raptors as they soar over the forest. If you are staying in Şavşat do bear in mind that while it may be nice and sunny around Ardahan, the weather can be awful on the west side of the Çam Geçidi, even in July and August. The road over the pass is a dirt track, and descending in a storm and/or rain can be an extremely frightening experience - a 1,000 metre descent on a muddy skidpan!

Birds

Common Cranes breed in the river valleys and small wetlands, with several pairs within a few kilometres of Ardahan. It is quite possible that Demoiselle Cranes occur in the area, although there are no records to date. White Storks are very common and Black Storks have been noted in July and August, suggesting possible breeding in the area. Redshanks and Lapwings are common, and many species of waders can be seen on passage, particularly in some of the inlets on the east shore of Çıldır Gölü. Several pairs of Citrine Wagtail are to be found in the area. Both Dalmatian and White Pelicans occur on Çıldır Gölü, usually in small numbers and Ruddy Shelduck and Ferruginous Duck can be seen among low numbers of other wildfowl. Velvet Scoter have been recorded at Çıldır in the summer months. Quail, Isabelline Wheatear, Whinchat and Lesser Grey Shrike are all quite common.

Raptors occur here in almost unbelievable numbers, both on passage, and seemingly, breeding. Observations in late July 1994, yielded dense populations of both Booted and Lesser Spotted Eagles in

Dalmatian Pelican

the more open forested areas to the east and southeast of Ardahan. Buzzard, Long-legged Buzzard and Hobby are also quite common, and in the marshy areas and steppe, both Montagu's and Marsh Harriers are common. Black Kite, Short-toed Eagle, Egyptian Vulture, Spotted Eagle, Kestrel and Lanner have all been noted over the forests and steppe mosaic east of Ardahan in July. Look out for Imperial Eagles between Ardahan and Çıldır Gölü, especially around the lake itself, and Griffon Vultures are not uncommon in the area, notably near the gorge a few kilometres from Çıldır. In the autumn, these species are joined by large numbers of Buzzards of the *vulpinus* race (birds of the chocolate-brown morph also occur here), and smaller numbers of various raptors including Steppe Eagle, Pallid Harrier and Saker. The whole region is a very exciting place to birdwatch, especially in migration periods when more or less anything can turn up. White-winged Lark has been recorded from the shores of Çıldır Gölü.

Other Wildlife

Wolves still occur in the area but are unlikely to be seen. The floral displays on the hillside steppes are superb and there are many rare and beautiful plant 'gems' to be found in the gorges.

Western Anatolia

This region is one of the lower altitude areas of Turkey, with large areas under agriculture. However, there are several significant mountain areas, with Uludağ reaching 2403m. The rocky coastline of the Aegean Sea to the west contrasts with the beaches of the Marmara Sea to the north.

Manyas Gölü

Manyas or Kuş Gölü was Turkey's first nature reserve, and its title, Kuş Cenneti (Bird Paradise) aptly describes the protected area. Unfortunately, the rest of the lake has deteriorated but the reserve is still spectacular with its breeding colonies of herons, spoonbills, cormorants and Dalmatian Pelicans.

Location

Kuş Cenetti is well signposted, and lies just off the main Bandirma to Balıkeşir road about 14km southeast of Bandirma. A modest entrance fee is payable on entry, and access is limited to the paths which lead to the tower hide. This is no great problem though, since the hide affords excellent views over the reeds, scrub and the heronry.

There are no buses to the reserve itself, although you can get a dolmuş to stop at the junction off the main road and walk from there.

Accommodation

There are several hotels in Bandirma, and a couple in Karacabey. Karacabey is an excellent base for the area, being within easy reach of Manyas, Uludağ and the Kocaçay Delta. There are also campsites all

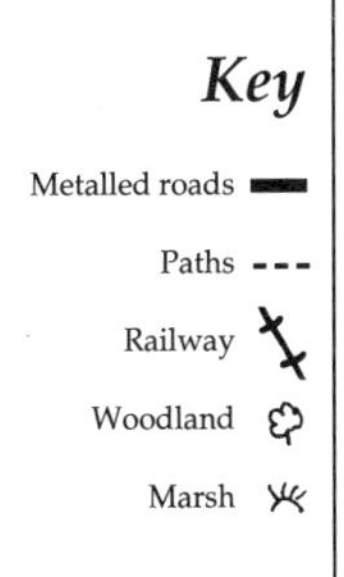

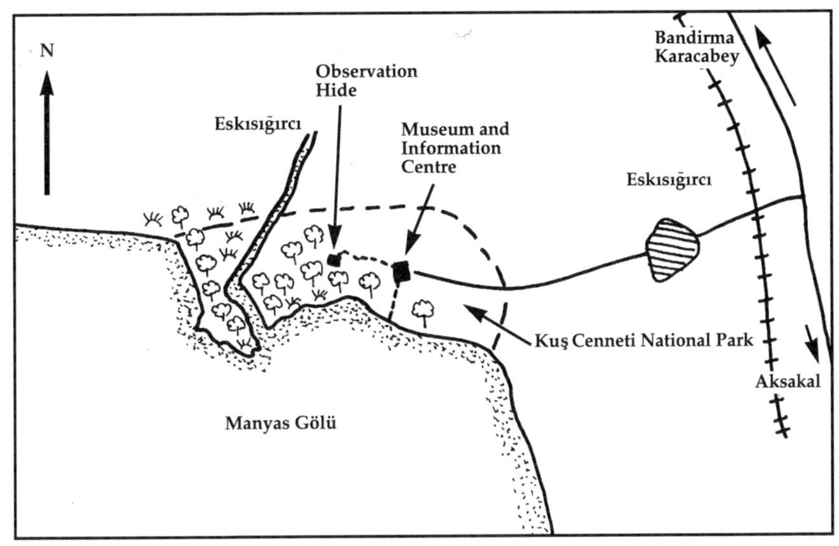

along the coast, but none within easy reach of the lake.

Strategy

The best times to visit are either during the breeding season or the migration season. Over 200 species have been recorded at the site, and 90 or so breed. The reserve is open for most of the day, but there is no access in the early morning. Outside the reserve the rest of the lake is fairly open, and most of the larger species can be seen without going inside.

Birds

The willows of the heronry hold breeding Grey and Night Herons, Little Egret, Spoonbill and Cormorant, and Dalmatian Pelicans breed on specially built platforms. The reedbeds hold Pygmy Cormorant, Purple Heron, Little Bittern, Penduline Tit and Great Reed Warbler. The scrub has breeding Great Spotted Cuckoo, Scops and Little Owls and Olivaceous and Savi's Warblers. Other species present all summer include Marsh Harrier, Lesser Kestrel, Spur-winged Plover, Roller, Bee-eater and Rufous Bush Robin. Large flocks of Glossy Ibis, White Pelican and White Stork stop off on passage, as well as good numbers of Red-footed Falcons and Caspian and White-winged Black Terns.

Kocaçay Delta

The Koca River drains Apolyont Gölü, and where it enters the Marmara Sea, there is a superb area of woodland and coastal scrub. As well as this, there are also two lakes, Dalyan Gölü and Arapçiftliği, with significant reedbeds and this wide range of habitats provides refuge for a good number of species, and as it is seldom visited by birdwatchers, there is much still to discover there.

Location

The delta is about 30km north of the town of Karacabey and can be reached by taking the road north out of town and following it to the coast. Just before the village of Yeniköy Plaj on the coast, take a track to the right which runs between the scrub and the beach. This track

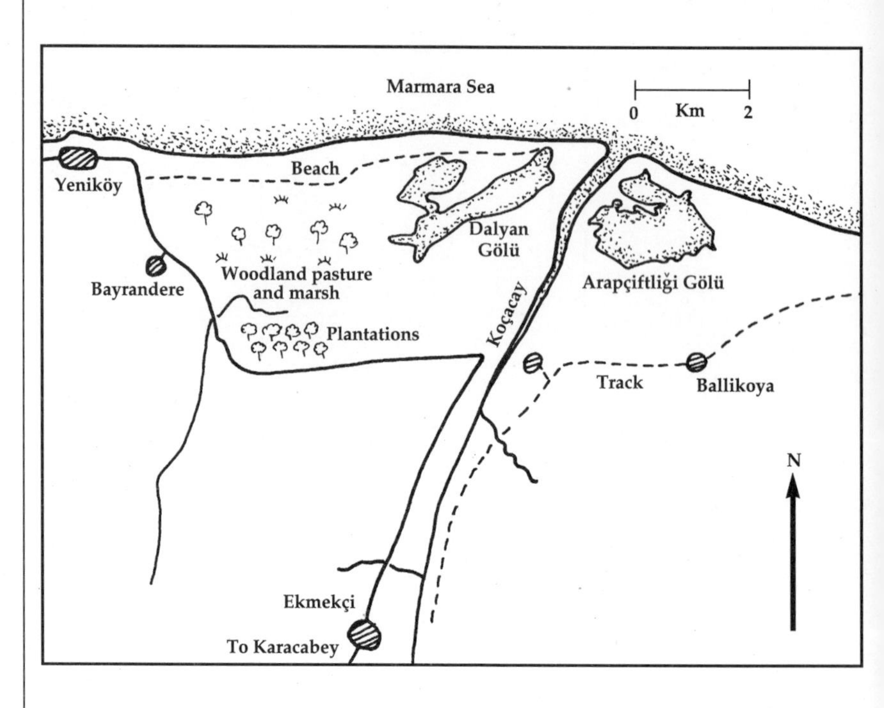

will take you through the scrub and as far as Dalyan Gölü. There are dolmuşes from Karacabey to Yeniköy Plaj, and the area is easily worked on foot.

Accommodation

There are a couple of hotels in Karacabey and camping along the beach at Yeniköy Plaj. There are restaurants in Karacabey, but you should take provisions with you during the day.

Strategy

The whole area offers excellent birding and is little explored, and so is worth a couple of days or more. It is possible to see a wide range of species in just half a day, and so can be combined with Manyas Gölü for those short of time. The best time to visit is between April and June.

The scrub on the hillsides near Boğaz village on the road to Yeniköy is worth checking for *Sylvia* warblers. The flooded scrub along the coastal strip should be thoroughly explored. Keep a look out for Lesser Spotted Eagle and Honey Buzzard flying overhead. The lakes are fairly devoid of birds, but are still worth a look.

Birds

The commonest birds in the coastal scrub during the breeding season are Red-backed and Masked Shrikes and Nightingale. Olive-tree Warbler has been recorded, but Olivaceous is much more common. There are several woodpeckers in the area, and White-backed Woodpecker has been seen as well as the more regular Syrian, Lesser Spotted and Green. The woods in the area have breeding Lesser Spotted Eagle and Honey Buzzard, and these can be seen overflying the scrub, along with Black and White Storks. Other species present include Bee-eater, Roller, Golden Oriole, Cetti's Warbler and Cirl

Krüpers Nuthatch

Bunting. The lakesides, beach and reedbeds have Ruddy Shelduck, Stone-curlew, Kentish Plover, Collared Pratincole, Tawny Pipit and Great Reed Warbler. River Warbler has also been recorded here.

The scrub near Boğaz has Nightingale, Woodchat Shrike, Rüppell's Warbler and Ortolan and Cirl Buntings. Pygmy Cormorants and Eleonora's Falcons have been seen flying over the area.

Other wildlife

Several pairs of Otter live in the delta area and Jackals are occasionally seen.

Uludağ

Uludağ is the highest mountain in Western Anatolia, and has a good range of high-altitude species not found elsewhere in the region. It is a national park, and is heavily visited by tourists.

Location

The mountain is located just south of the town of Bursa, formerly the capital of the Ottoman Empire, and can be reached by taking the road just west of Bursa off the Karacabey road, which is signposted to the park. A long winding road climbs through thick beech forest and then coniferous forest towards the summit. The road is about 15km long so it may be advisable to get a dolmuş from Bursa. There is also a cable car to near the top, and a dolmuş from there to the summit hotel.

Accommodation

There is a wide range of accommodation in the area, and new hotels are being built. There are cheaper hotels in Bursa, and several more expensive ones at the mountain. The same is true of restaurants.

Strategy

The best time to visit is spring and summer, but there can be several metres of snow on the summit area even as late as June.

Birds

The pine woods along the road hold Krüper's Nuthatch, and Tengmalm's Owl has also been recorded here. The higher areas hold Shore Lark, Water Pipit, Alpine Accentor, Rock and Blue Rock Thrushes, Chough, Alpine Chough and Red-fronted Serin. Lammergeier is regularly recorded, but is difficult to find, while Griffon and Egyptian Vultures and Golden Eagle are more common. On the lower slopes Red-rumped Swallows, Alpine Swifts and Red-backed Shrikes are common. Uludağ is one of the few currently known sites in Turkey for Pallid Swift, with small numbers summering here.

Bafa Gölü

This large lake is situated on the southern edge of the Büyük Menderes river valley, and was originally a bay of the Mediterranean sea. The northern shores are flat and marshy, the rest are hilly, or even mountainous (in the south).

Key

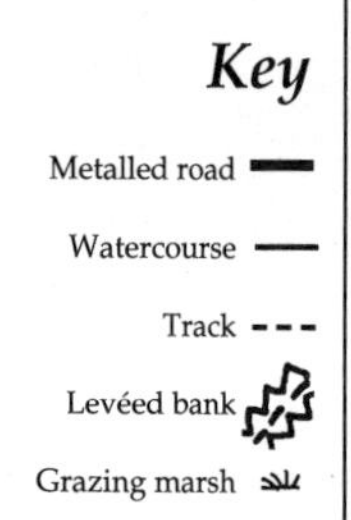

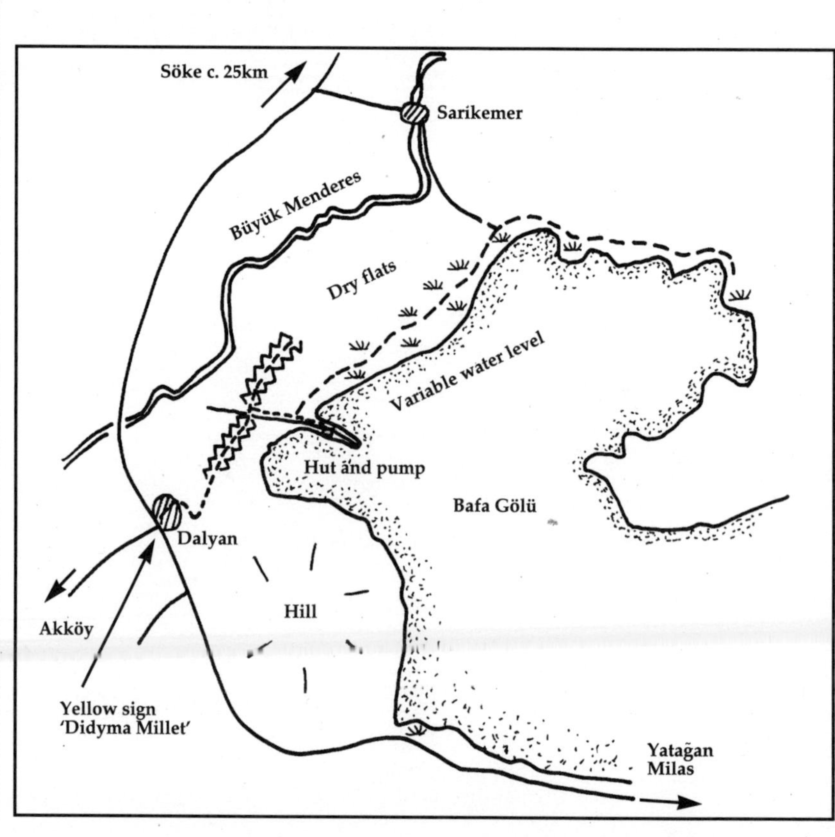

Location

Bafa is about 150km south of İzmir on the İzmir to Milas road (route 525). This road skirts the western shore, with the lake nearly always in view. Most of the lake is of little ornithological interest, the majority of the wetland species occurring along the northern shore. To reach this area, take the eastbound track, at a crossroads situated at the southern end of the flat Menderes valley floor, just before the road rises into some low hills. The westbound road is signposted to Didyma (there is a stork's nest on a pylon next to the junction). Follow the eastbound track, which goes along the base of the hills, and eventually forms the top of a levee. As the track leaves the base of the hills, the first track off to the right leads down to a small building, which, depending on water levels, is only a few hundred metres from the shore. Beware of the dogs around the building!

Accommodation

There is a recently opened campsite on the lakeshore and one at the Turgut Hotel, both at the southeastern end of the lake. There may be some accommodation available in tea houses in nearby villages. There is a motel, the Tufan Efes Moteli, near the town of Selçuk, and several hotels in the town itself, which is about 60km north of Bafa. The tourist resort of Kuşadası offers a wide variety of accommodation, and is a little closer. One of the seven wonders of the ancient world, the Temple of Diana at Ephesus, lies between Selçuk and Kuşadası.

Strategy

Bafa provides good birdwatching for most of the year, with mid-summer being the quietest period. The best areas are the marshy 'flats' to the east of the track leading to the shore, and the end of the causeway leading into the lake from the 'building with dogs'. Dalmatian Pelicans often sit on the very end of the causeway, or can be seen on the water nearby. The rest of the lake is worth checking for White-tailed Eagle.

Birds

Along with Manyas Gölü, Bafa is the best place in Turkey to see one of the Western Palearctic's most endangered species, the Dalmatian Pelican. The birds can be seen at most times of the year, although they are more numerous in winter, with up to 400 being recorded. They seem to commute regularly to the Büyük Menderes delta, in particular to Karine Gölü, so they are not present throughout the day. White Pelican has also been noted on occasion.

Spring and autumn passage brings good numbers of Spoonbill, Little Egret, Squacco, Night and Purple Herons, Little Bittern and Glossy Ibis. Large numbers of Whiskered and White-winged Black Terns pass through, while flocks of up to 500 Mediterranean Gulls occur in spring. Collared Pratincole, Black-winged Stilt and Spur-winged Plover are all common, and good numbers of passage waders can be found on the muddy fringes. Wildfowl are almost absent during the summer months, but large concentrations occur in winter.

White-tailed Eagles probably breed in the area, and can be seen hunting over the marshes, especially in autumn. Short-toed Eagles and

Long-legged Buzzards breed in the surrounding hills, while in spring, small flocks of Red-footed Falcons, Hobbies and Eleonora's Falcons 'hawk' over the northern shore. Calandra Lark, Yellow Wagtail and Tawny Pipit frequent the areas of Salicornia on the lake fringes.

The hills to the south of Bafa hold good populations of Krüper's Nuthatch, Sombre Tit and Cretzschmar's Bunting, and Bonelli's Eagle may well occur in the area.

Other wildlife

The flora in the hills around Bafa is exceptional in April when the olive groves are often carpeted in a multi-coloured display of anemones, with both *Anemone pavonina* and *A. coronaria* being very common. Orchids are also well represented with Bug, Holy, Anatolian, Italian, Monkey, Yellow Bee, and Tongue Orchids, as well as *Ophrys reinholdii* and *O. mammosa*, all in flower in late April and early May. Agama Lizards are common and Glass Snake and Snake-eyed Lizard inhabit the maquis-covered hills. The spectacular Two-tailed Pasha is one of the many species of butterfly found in the vicinity. The mountains to the south-east of the lake hold a small population of Brown Bear.

Dalyan

Dalyan is a useful base from which to explore the amazingly diverse habitats to be found in the far southwest of Turkey. Within a few kilometres of the town there are lakes, freshwater and estuarine marshes, phrygana (degraded maquis scrub) covered limestone hillsides, *Pinus brutii* forests, *Liquidamber orientalis* forests, rocky and sandy shores, high mountains, and a bewildering variety of agricultural crops. There is also plenty to interest other tourists, and as such Dalyan is an excellent place to combine a relaxed holiday with birdwatching.

Location

Dalyan is situated on the River Calibis which links Köyceğiz Gölü to the sea. The town is signposted as Kaunos (the ruins) from the main Muğla to Fethiye coastal road (route 400). The easiest route is to take the road from Ortaca, however the minor road to Dalyan, which meets the coastal road about halfway between Köyceğiz and Ortaca, is rather beautiful as it runs past the *Liquidamber* forests.

Local public transport is excellent, with regular dolmuş services from all nearby towns. There are several travel agencies in Dalyan which can arrange transport further afield. There is a large boat co-operative in Dalyan with boats hired out (with 'driver') for around £20-30 per day, or negotiable for shorter periods. This is an ideal way to birdwatch in the marshes, or around the lake, and some of the boatmen are familiar with the wants of birdwatchers. For those wanting to visit the ruins across the river, there is a row boat ferry costing a few pence. Car hire is expensive locally and it is often cheaper to hire a taxi for a day trip to the hills. Dalyan is about 30 minutes drive from the international airport at Dalaman.

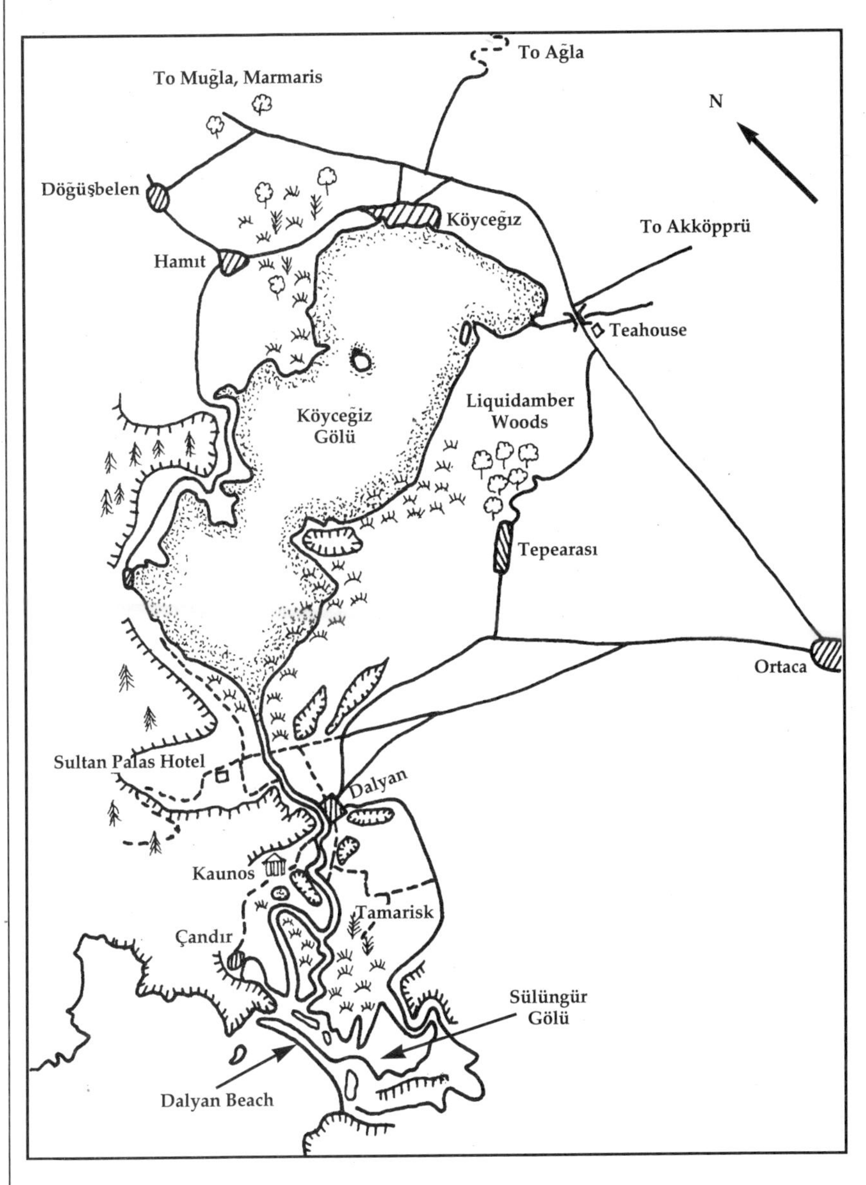

Accommodation

Dalyan, not surprisingly, is full of hotels and pensions. Although more expensive than much of Turkey, prices are still reasonable, ranging from £5 to £6 for pensions, to around £30 a night for the better hotels. For a touch of luxury, try the Sultan Palas Hotel across the river, or in Dalyan itself, the Kaunos Hotel on the waterfront has a friendly atmosphere and is cheaper than most. Restaurants are plentiful, and reasonably priced. Seafood is a local speciality with excellent squid, octopus and various fish dishes available.

Strategy

The area provides decent birdwatching all the year round. However, the most interesting periods are April to June, and August to November. The normal tourist season is in full swing from late June to September, so this should be borne in mind as boat traffic is quite heavy through the marshes to the beach and disturbance levels are quite high. Because of Dalyan's location it is rarely visited by birdwatchers travelling around the usual circuit. However, this region does provide one of the best chances of finding Smyrna Kingfisher in Turkey. The best area to search for this species is in the environs of Hamit, a small village near the lakeshore about 6km northwest of Köyceğiz. Nearer Dalyan, the most likely areas are the large reedbeds on the eastern shore of Köyceğiz Gölü, and the arable land between the marshes and hills just south of Dalyan, particularly around the 'obvious' large area of Tamarisk about a kilometre south of the town. If several days are available, visits should be made to the *Liquidamber* forests near Tepearası (and perhaps to those just north of Köyceğiz) and to the hills around Ağla and Akköprü. Locally the ruins at Kaunos provide excellent birdwatching, as do the small areas of marsh either side of Dalyan. During migration periods, look for cotton fields, which are periodically inundated with water, as these attract both waders and passerines.

Birds

All three species of kingfisher occur in the area. There are one or two pairs of Smyrna Kingfisher breeding locally, and several pairs of Lesser Pied Kingfisher, mostly around Köyceğiz Gölü. Although scarce during the breeding season, Kingfishers are also present, and their numbers build up to several hundred by late autumn. Other waterbirds of interest include various heron species, with noticeable movements of Night and Squacco Herons and Little Bitterns in spring. Pygmy Cormorants are reasonably common from May to October.

The main lake is strangely devoid of waterfowl except in winter, however there is a good passage of Whiskered and White-winged Black Terns in spring and autumn. White Storks are common and there is a breeding colony in the pines on the edge of town. Raptor migration is quite good in the area and a good watchpoint is from the ruins at Kaunos, or if you prefer refreshment at the same time, from the waterfront outside the Kaunos Hotel. Large numbers of Red-footed Falcons, Hobbies and Eleonora's Falcons pass through, and some of the latter two species may stay to breed. Long-legged Buzzards, Short-toed Eagles and Lesser Kestrels all breed locally. All four species of harrier are recorded on passage, with Hen Harrier wintering, as does White-tailed Eagle, which can also be seen occasionally during the summer. Other raptors to look out for, some of which may breed, include Peregrine, Goshawk, Sparrowhawk and Lesser Spotted, Booted, Golden and Bonelli's Eagles. Scops, Little, Tawny and Long-eared Owls breed locally and Eagle Owl probably does so as well. Scops Owls can be heard and often seen around the town itself (check the muezzin's platform on the mosque in the town centre).

Common breeding species in the phrygana around Dalyan, and the ruins, include Roller, Red-rumped Swallow, Rufous Bush Robin, Black-eared Wheatear, Olivaceous and Rüppell's Warblers, Rock Nuthatch and Black-headed Bunting. Sardinian and Orphean Warblers also breed but are quite scarce. Chukars are frequently heard at dawn or dusk, but are difficult to get more than a glimpse of. The tamarisks have extraordinarily dense populations of Olivaceous and Cetti's Warblers, and where there are willows or other larger trees among the tamarisks, look for the beautiful hanging nests of Penduline Tit. The marshy areas have Great Reed Warblers and Yellow Wagtails breeding, and during migration periods attract rails, snipe and an assortment of passerines.

The pine forests on the hills around Ağla and Akköprü hold several species of woodpecker, Sombre Tit, Dipper, Grey Wagtail, Redstart (the *samamisicus* race with the white wing bar) and Krüper's Nuthatches are very common. The barren slopes above and below the forests are good for Rüppell's Warbler and Cretzschmar's Bunting, and less commonly, Nightjar and Woodlark.

The *Liquidamber* forests are endemic to this region of Turkey, and are extremely localised within it. Their appeal is as much aesthetic as for birds. Breeding species in the forest, and the meadow edges, include Middle Spotted, Lesser Spotted and Syrian Woodpeckers, Roller, Hoopoe and Red-backed and Masked Shrikes. During migration, large falls of passerines, in particular the *Ficedula* flycatchers occur in these woods.

Scarcer species regularly recorded in the Dalyan area include Dalmatian Pelican, Shag, Glossy Ibis, Great White Egret, Spotted Crake, Slender-billed Gull, Red-throated Pipit and Fan-tailed Warbler.

Other Wildlife

Dalyan is famous for its turtles, with one of the few breeding sites of the Loggerhead Turtle *Caretta caretta* in Turkey. The area was designated a Specially Protected Area in 1988, and the breeding beaches are protected by soldiers during the breeding season. The Nile Soft-shelled Turtle *Trionyx triunguis* occurs in the freshwater parts of the wetland complex. This last species is to be found only at Dalaman and here, and there are now probably under 100 individuals left.

Butterflies are plentiful in the area with False Apollo and Eastern Festoon common, especially in the hills in May, and colonies of Large Tortoiseshell around Dalyan. Europe's largest butterfly, the Two-tailed Pasha, can sometimes be seen along the coast here, notably at Ekincik, and the smallest, the Grass Jewel, can be found in the Dalaman Gorge at Akköprü. The flora is exceptionally varied with a wide range ooo orchid species. One of the most obvious plants of the area is the spectacular Dragon Arum *Oracunculus vulgaris* with its huge purple bracts and repulsive smell of rotting corpses.

SELECTED SPECIES

Pygmy Cormorant A decidedly local species through much of its range, Pygmy Cormorants seem to be increasing in numbers in Turkey. They frequent wetlands with a reasonable amount of cover, particularly those with large reedbeds and/or wooded fringes. Good populations occur at Ereğli (p20), Sultan marshes (p22) and at most of the coastal wetlands, particularly in the west. Small numbers also occur in the Van region (p56) and in the river valleys to the north of there. They are even more populous in winter, again predominantly in the west.

White Pelican The commoner of the two pelicans, they are still very local in occurrence and can regularly be found in numbers at only a few wetlands. They may breed at Manyas Gölü (p82) and Sultan marshes (p22) and there is a recently discovered colony at Seyfe Gölü. Other sites where they are normally encountered are Bafa Gölü (p86) , Ereğli marshes (p20) and the Göksu delta (p37). There are a number of records of quite large flocks from Bendimahi (p63) on the northeast corner of Van Gölü, and smaller numbers from Haçlı Gölü to the northwest, but these do not seem to be regular occurences and may well relate to populations on Daryacheh-ye-Reza'Iyeh (Lake Urmiah) over the nearby Iranian border.

Dalmatian Pelican A globally threatened species. Turkey holds an important breeding population of this species and a significant proportion of the world population winters in this country. They are known to breed at Manyas Gölü (p82) in small numbers (on a man-made platform within the sanctuary, viewable from a hide), in larger numbers in the Bafa Gölü and Büyük Menderes delta complex (p86), and at Çamaltı Tuzlası near İzmir. In winter and spring up to 400 can be found in the Bafa Gölü and Büyük Menderes delta complex (p86). They also breed in the Kizilirmak delta (p49) and formerly in the Meriç Deltası and İznik Gölü. There are records involving small numbers or singles from many parts of the central plateau and the east.

Black Stork An uncommon summer visitor and passage migrant. Black Storks are recorded in surprisingly low numbers at the main migration watchpoints, although they can be seen anywhere in central, western and northeastern Turkey during spring or autumn. Turkey represents a southwestern outpost of their breeding distribution. Look for them in the well-forested valleys in the hills between Gerede and Ankara where they breed in reasonable numbers - they can almost be guaranteed to be seen from the road between Gerede and Kızılcahamam, and in the Soğuksu Milli Park (p16). Their distribution eastwards from this area is little known but they have been noted in suitable breeding habitat in June as far east as Elaziğ and Malatya, and the Çoruh valley near Yusufeli..

Bald Ibis The story of the Bald Ibis colony at Birecik (p30), the only known breeding colony in the Middle East, is a sad one. The wild population became extinct in 1989, although there are around sixty free-flying birds still in residence, the result of captive breeding at the station by the wadi entrance just north of the town. A rapid decline started in the fifties when a bridge was built across the Euphrates. The bridge precipitated a rapid growth in the town's population and also destroyed the jobs of the ferrymen who held the bird sacred and who celebrated the birds' return every year with a festival, with the result that the Ibis was no longer associated with the well-being of the town. Much worse was to follow with heavy spraying of insecticides to eradicate malaria, the use of pesticides on newly reclaimed wetlands in the area and the extensive use of 'prophylactic' chemicals to combat a vast locust swarm threatening south-east Turkey. The combination of these events was too much for the Bald Ibis which is particularly sensitive to pollution and breeding failed year after year until the population aged and dwindled. The then World Wildlife Fund set up a project to try to save the birds, but encountered many problems, including the tipping of rubbish onto the breeding ledges and buildings appearing right above the colony. The long term effects of the pollution meant that the colony was never able to recover, and although the captive breeding pogramme succeeded in raising some young, the lack of knowledge of their breeding ecology ultimately led to failure.

Greater Flamingo The breeding status of this species is little understood in Turkey. Large colonies have been found in the Tuz region on at least two occasions in the last twenty years but it is not known whether they are a regular breeder. The vast area of soda lake and salt steppe in the area make observations very difficult, and only one thing is certain - that there are regularly large numbers of Greater Flamingo in the area, notably at Seyfe Gölü (where up to 30,000 have been seen, and they have bred), Kulu Gölü (p18) and at several small wetlands near Cihanbeyli. There can also be large concentrations at other central plateau wetlands, such as Ereğli marshes (p20) (where they were found breeding in 1992) and Sultan marshes (p22), and at Burdur Gölü (p26). Elsewhere, they do not occur in such phenomenal numbers, although they may be found at many wetlands in both the west and the east. Records from around Van, as with White Pelican, probably relate to birds from the Iranian populations.

Marbled Duck A rare and declining species throughout its range and seemingly becoming much rarer in Turkey. Marbled Ducks are mainly summer visitors although they do overwinter in small numbers. They breed in very low numbers in a few wetlands on the central plateau such as Ereğli marshes (p20) and Sultan marshes (p22) and possibly (formerly?) Hotamiş. In the southern coastlands they are reasonably common at the Göksu Delta (p37), and there is also a population in the Seyhan/Ceyhan delta. In the east they have been seen with young in some of the lagoons fringing Van Gölü (p56), and occur in montane steppe river valleys in the region.

White-headed Duck Turkey is a very important country for this globally threatened species. In winter up to 90% of the world population congregate on Burdur Gölü (p26) with a maximum winter count of c.11,000. They also occur in small numbers on nearby lakes such as Yaraslı Gölü and Karataş Gölü (formerly a higher proportion but recently the birds have prefered Burdur). In 1992 Burdur City Council banned hunting of this species and appears to want to take an active part in the conservation of White-headed Duck. Turkey also holds a significant breeding population with small numbers found at wetlands in the central plateau such as Kulu Gölü (p18), Ereğli marshes (p20) and Sultan marshes (p22), and at a number of small wetlands on the shores of Van Gölü (p56). They seem to prefer shallow warm water with plenty of emergent vegetation for their breeding territories, often lagoons on the shore of a larger body of water, and it is in these more open areas that post-breeding and passage congregations occur. Kulu Gölü (p18) and Arın Gölü (p70) often hold more than 500 birds each in late summer and autumn.

Velvet Scoter A somewhat unlikely Turkish breeding species as the nearest part of the main range is some 2000 miles to the north; one can only guess at the reaction of the first ornithologists who found them!. They are known to breed at several high level lakes in the northeast of the country. The easiest and most spectacular place to find them is at the crater lake in Nemrut Dağı (p72) on the west shore of Van Gölü (p76). They are also found at Çıldır Gölü (p78) and Balık Gölü. They can occasionally be found around Van Gölü itself in summer months along with other northern wildfowl such as Goldeneye and Long-tailed Duck.

White-tailed Eagle A very rare resident and winter visitor with perhaps less than 10 pairs breeding in the whole of Turkey. Outside the breeding season they have been recorded from a wide variety of areas but in recent years most records of breeding have come from the southwest of the country. One or two pairs can be regularly found around Bafa Gölü and the Büyük Menderes delta (p86), a pair at Çamaltı Tuzlası near İ zmir, and another pair is resident around Dalyan (p88). It is possible that there are other pairs in the wild rocky coasts of this region. Elsewhere they do not seem to occur in the 'lake district' which, especially around Beyşehir, would appear to be suitable habitat, or from the northeast where they might also be expected.

Lammergeier A widespread resident throughout Turkey although the vast territories of each pair mean that the total number of breeding pairs probably is not in excess of five hundred. They can be found in any of the high mountain areas especially where there are plenty of gorges and cliffs, and also in sparsely populated hill country such as around Kızı lcahamam. The best areas to look for Lammergeier are in the Kaçkar Dağları (p50) in the eastern Pontics, the Aladağlar (p39) and Bolkar massifs of the Toros mountains, Uludağ (p85), and the Çoruh river gorges from İspir (p46) to Artvin.

Black Vulture This species occurs sparingly over much of the hilly and mountainous areas of the country. They are found along the Pontic Alps and along the Toros Mountains, with their stronghold in the Köroğlu Dağları and the Karagüney Dağı roughly in a triangle between Gerede, Ankara and Corum. Here they seem to favour the sub-Pontic and Anatolian pine forests where *Pinus brutia* and *P. pallasiana* are dominant, habitat in fact not dissimilar to their favoured habitats in Extremadura, Spain.

Levant Sparrowhawk A very scarce breeding species known with certainty only from the northwest of the country. On migration they can be seen in considerable numbers in autumn at the Bosphorus with smaller numbers to the east at İskenderun/ Belen pass area and in the northeast in the Borçka (p53) and Arhavi region. An example of the relative numbers for two of these; 2,900 at the Bosphorus in autumn 1972 and 290 at Arhavi/Borçka in autumn 1976.

***Aquila* spp.** The various *Aquila* species found in Turkey can best be summarised as follows.
Golden Eagle; reasonably widespread resident especially in the mountainous areas.
Imperial Eagle; uncommon resident in the central plateau and the east, most likely to be encountered around the fringes of the central plateau and in the mountain valleys between Van, the northeast border and Erzurum. More frequent on migration, especially in the east.
Lesser Spotted Eagle; breeds in low numbers mostly in the north and west with a significant population around Ardahan (p78). The Kocaçay delta (p83) is a good area to look for this species. Large numbers pass through the Bosphorus and the eastern migration corridors.
Spotted Eagle; uncommon winter visitor particularly to the western and southern coastlands, also noted on passage although much less common than the above species.
Steppe Eagle; Passage migrant, reasonably frequent in the east becoming steadily rarer to the west.

Bonelli's Eagle An uncommon breeding species, possibly much rarer than previously thought. Known with certainty from a very few areas but may be more widespread in parts of south and west Turkey which have been little visited by ornithologists. Probably the easiest place to find them is at Halfeti (near Birecik) (p30) where they breed on the cliffs across the river Euphrates from the village.

Eleonora's Falcon A rare summer visitor known to breed in only a few localities in the Sea of Marmara and around the southwestern coasts. Small groups of up to 15 birds can be found in late spring around the coast near Dalyan (p88) and they are regularly recorded from the southern coastlands. Interestingly they can quite often be found well inland along the southern flanks of the Toros. Anyone coming across an all dark falcon in the southeast or eastern regions should be aware of the possibility of Sooty Falcon occurring in the area.

Lanner/Saker Both of these large falcons are undoubtably uncommon, but their true status in Turkey is unclear as many old records may well be erroneous, because until recently there was little good literature on identifying these difficult birds. Both are scattered throughout the central and eastern regions, with Saker showing a preference for the open steppelands of the central plateau and the eastern montane steppelands, and Lanner preferring more rugged mountain country such as the Toros. However they can occur in the same localities, especially during migration. Lanner can be found right down to the southwestern coastlands. One of the best areas to look for Saker is in the high mountain valleys to the north of Van Gölü (p56).

Caucasian Black Grouse Turkey holds the only Western Palearctic population of this species outside of the Caucasus. They are a rare resident (altitudinal migrant?) in the northeastern Black Sea coastlands. The majority of the records come from one site, Sivrikaya (p43), largely because this is probably one of the easiest points of access to their high mountain range. They frequent alpine rhododendron scrub usually between 2200m and 3000m. Their range seems to be constrained largely by the availability of this habitat which is more or less exclusively found on the northern slopes (and north-facing slopes south of the main mountain ridge) of the eastern Pontics. Habitat degradation has occured around Sivrikaya, however on a recent trip into the Kaçkar Dağları (p50), which would seem likely to be their stronghold, there appeared to be plenty of good habitat available. The only other area where they are likely to occur is on the ridge tops accessable from the Çam Geçidi (R 965), east of Şavşat (p54).

Caspian Snowcock A resident high alpine species with a wide distribution over central and eastern Turkey. They are usually found above 2000m, and from mid-summer are often difficult to locate below 3000m, preferring to stay on the high alpine scree up to and above the snowline. This preference for the highest peaks (particularly snow-covered and craggy ones) make them difficult to find, but their far-carrying calls can often be heard echoing around cirques and ridges in the early mornings. They are reasonably common in the Aladağlar (Toros) (p39) and in the Kaçkar Dağları (Pontics) (p50), and have been recorded from most suitable mountain areas including the Cilo/Sat Dağları near Hakkari.

See See A bird of arid areas confined to the southeastern region in the areas bordering Syria. Within this area they can be found where there is suitable rocky arid habitat. Only regularly recorded from the area around Birecik and Halfeti (p30), and further east near Cizre and İdil (p35).

Black Francolin A rare and local resident in the southern coastlands and southeast. They inhabit scrubby areas often in the vicinity of water such as dune systems or riverbanks. In the last century they seem to have occured over a much wider area, but now they are more or less restricted to the deltas of the Göksu (p37) and the Seyhan/Ceyhan where they are still reasonably common. They are also known from the Tigris valley at Cizre (p35) and presumably may be found in other similar habitat in the southeast.

Common Pheasant Included in this account as there is some debate as to the exact status of this species in Turkey. It is possible that the westernmost extent of the range of wild birds may reach Turkey so careful notes should be taken regarding the 'race' observed, particularly in the northeast. Records from the west of Turkey will almost certainly be from introduced stock.

Crakes Five species of crake are recorded in Turkey, mostly as passage migrants. Water Rail is a reasonably widespread breeding species. Baillon's, Little and Spotted Crakes are infrequently recorded, usually in migration periods, although they may also breed. These last three are likely to be much commoner than records suggest, and if a suitably secluded spot in a wetland can be found in spring or autumn then it is quite possible to see all three at a single visit. Corncrake, is a rarely recorded passage migrant, noted from a wide variety of habitats and locations.

Purple Gallinule A rare resident in small numbers at the Göksu Delta (p37) where it is quite easy to locate. Nearly all recent Turkish records come from this site, although they have recently been found breeding in the Seyhan/Ceyhan delta at Akyatan Gölü. The species was recorded as quite common at Amık Gölü (near Antakya) until it was drained in the 1960's. They are also occasionally recorded from lakes on the central plateau, but while there is plenty of suitable reedbed habitat, the paucity of records does not suggest a secure breeding population.

Demoiselle Crane An extremely rare and local summer visitor breeding in the northeast of Turkey. During the breeding season they frequent large river valleys in the eastern montane steppelands where traditional agricultural practices are still maintained. Formerly more widespread in the northeast, they currently breed in the Murat Nehri valley near Bulanık (p75) and may do so as far west as Elazığ. They have been recorded in the breeding season in the valleys around Ağrı (especially near Söylemez), Kars and Doğubeyazıt (p68). On passage they could be seen anywhere in the country although records are infrequent.

Little Bustard Status uncertain. There are very few records in any season and although they are likely to breed there is only a couple of recent observations from this season, near Ceylanpınar (southeast) and near Burhanlı (southwest of Ankara). There are more records from the winter months indicating that there may be an influx in this period.

Great Bustard An uncommon resident of steppelands and extensive agriculture. May also be a partial migrant since eastern areas are snowbound in winter. The distribution is centred on three main areas: The central plateau (particularly in the vicinity of Tuz Gölü), around Ceylanpınar in the southeast (where they are said to be much commoner in winter), and in the eastern region from the Iranian border westwards to Muş. The latter probably holds a reasonably large breeding population as they have been recorded in large groups from many areas, notably east of Erçek (p60), the Göldüzü area, the Murat Nehri valley around Bulanık (p75) and Patnos, and the same valley further west near Muş. There is a good deal of suitable habitat in the region, much of it far from road access, and drives along tracks such as those between Ahlat and Nazık should yield sightings of this species. Outside these areas they are rare.

Cream-coloured Courser An infrequently recorded species of the arid southeastern areas from Gaziantep east to Cizre (p35). Most records presumably refer to extralimital birds even in the summer months, however it is quite possible that they breed in this region in some years.

Black-winged Pratincole A passage migrant which is usually very scarce although they can appear in good numbers in some years. While they have been recorded from all regions, most of the observations have been from the eastern two-thirds of the country. They favour the wetlands of the central plateau and the river valleys to the north of Van Gölü (p56) where it is possible that they may breed occasionally.

Greater Sand Plover An uncommon and local summer visitor and passage migrant. Breeds in the inland salt steppes and sand and mudflats of the central plateau, and possibly the southeast. The best areas to look for this species are around Sultan marshes (p22) and Ereğli marshes (p20) as well as some of the smaller wetlands in the region such as Kulu Gölü (p18) and the remnants of Hotamış (p20). The bulk of the Turkish population probably breeds around Tuz Gölü, but looking for them in this area is not easy. On migration the wetlands of the southern coastlands are prime areas to search, particularly the Göksu delta (p37) and Seyhan/Ceyhan complex.

Spur-winged Plover Quite a common summer visitor to wetlands in the western two-thirds of the country. Even so it is curiously uncommon in many areas such as the 'Lake District' (p26) and only occurs in numbers at certain wetlands such as Ereğli marshes (p20) and Sultan marshes (p22), the Göksu and Seyhan/Ceyhan deltas (p37), Bafa Gölü and the Büyük Menderes delta (p86).

Red-wattled Lapwing First discovered in Turkey in 1983, this species appeared to be spreading up the Tigris valley where it was regularly recorded breeding on shingle islands in the river at Cizre (p35), near the Syria/Iraq border. However the recent problems in this area mean that little current information is available and the amount of fighting in the vicinity of Cizre in 1992 means that the current status of this species can at best be described as uncertain.

White-tailed Plover A very rare summer visitor and passage migrant. It has been found breeding on the central plateau at Hotamış (p20) but the (perhaps temporary?) disappearance of much of these wetlands must make it unlikely that they will breed at this site in the near future. They have also been recorded at other wetlands in the central plateau, southern coastlands, the southeast and east, but observations are sporadic. In 1992 they were observed displaying at Birecik (p30).

Migrant waders Large numbers of many species of wader using the eastern Mediterranean flyways use Turkey as a stopover. Any wetland can host a superb variety of waders, many of them less than familiar to most ornithologists from western Europe. Marsh Sandpipers, Broad-billed Sandpipers and Red-necked Phalaropes can sometimes be seen in good numbers, particularly in the east. Among the rarer species, Terek Sandpipers can be seen in small groups in the east, while Great Snipe are recorded from many wetlands and from the valleys in the montane steppelands of the northeast. Three great rarities should also be looked for; Caspian Plover (several records from the east and central plateau), Sociable Plover and Slender-billed Curlew.

Slender-billed Gull An uncommon resident and partial migrant. As a breeding species they are more or less confined to the central plateau. Breeding colonies have been found at Kulu Gölü (p18) (50 pairs in 1987, formerly more), Bolluk Gölü (declining), Karapınar Ovası (1000 pairs in 1971, 700 in 1972, apparently none since), Seyfe Gölü (2100 in 1971, not many now), Ereğli marshes (p20) and Sultan marshes (p22). This apparent decline is very marked and quite disturbing although the figures from 1971 may well not stand up to close examination. This species has been recorded from the Mediterranean and Aegean coasts mostly outside the breeding season, and is quite frequently observed around Van Gölü (p56) although its status is not clear in this area.

Audouin's Gull A very rare breeding species, probably resident although there may be local movements. Regularly recorded in numbers of up to twenty from around the Göksu delta (p37) area where they can often be found in Taşucu harbour or on the beach near the holiday homes at the west end of Akgöl. They breed further east along the coast and could be seen from any point along the coastal road in this region. They are sometimes found at other localities on the Mediterranean and Aegean coasts and occasionally on the western end of the Black Sea coast.

Terns Common and Little Terns are common summer visitors to much of the country, while Gull-billed Terns are much more localised with breeding colonies on a number of central Anatolian wetlands and eastern river valleys. Caspian Terns are a rare breeding species mostly in the Van basin and nearby river valleys, although sometimes they can be found on the central plateau. The marsh terns *(Chlidonias sp.)* can often be found migrating in vast numbers, but with the exception of Whiskered Tern are, at best, rare breeding visitors.

Black-bellied Sandgrouse A reasonably common resident in the eastern two-thirds of the country where they inhabit dry sparsely-vegetated regions. Although common, they are not easy to find. The best way to observe them is to find one of their regular drinking areas such as the gravel banks in the Euphrates about a kilometre north of Birecik (p30). Up to five hundred can be seen here in a morning. They are particularly common in the montane steppelands of the east.

Pin-tailed Sandgrouse Much less common than the above species. They are confined to the Syrian border area in the southeast where they are probably as common as Black-bellied, indeed a morning at the above-mentioned gravel islands may well yield up to five hundred of these as well. It is likely that they are quite common eastwards towards Ceylanpınar.

Laughing Dove Local resident, sometimes in considerable numbers especially in urban areas. A disjointed distribution concentrated around the İstanbul area in the northwest, and Birecik/Urfa/Gaziantep in the southeast (p30). They have also been recorded in smaller numbers from many other areas and it seems likely that they are expanding their range. Recently they have been found breeding in Samsun on the Black Sea Coast. The only region where they do not seem to have been recorded is in the east.

Striated Scops Owl A very rare summer visitor first discovered as a breeding species in 1982. They are known from only a few sites, all in the Euphrates valley just north of the Syrian border. The pair in the tea gardens at Birecik (p30) have now become famous, as much with the locals as visiting ornithologists and are reasonably easy to see, as are the Scops and Long-eared Owls which share the gardens! There is also at least one pair at Halfeti, and it is quite likely that other pairs could be found in any decent grove of trees in the area.

Brown Fish Owl The accidental capture (by a fisherman's hook!) of a single Brown Fish Owl near Adana in 1990 was the first substantiated record of this species in Turkey and was in habitat potentially suitable for breeding. During the last century specimens were collected from Mersin and Aydın and there are a few rather doubtful records from earlier this century. It seems likely that there is a small breeding population in Turkey but very little can be deduced from one recent record.

Tengmalm's Owl Another recent discovery in Turkey, this species has now been found in several widely spaced localities. They are presumed to be resident in pine forest at high altitude. So far they have been noted from Uludağ (p85), presumably an isolated population, and at a few locations from Kızılcahamam eastwards along the Pontic Alps. They are probably commonest in the northeast where there are extensive *Abies* and *Pinus* forests.

Little Swift A rare summer visitor to the a few localities in the southern coastlands and southeastern regions. Regularly observed at Birecik (p30), and particularly at Halfeti, where there are small breeding colonies. There have been a number of records from the Iskenderun area and as far east as Siirt. It is possible that this species is expanding into the region.

White-breasted Kingfisher A rare resident at the western extreme of its range. Formerly more widespread it is now found in a few localities in coastal lowlands and river valleys from the southern half of western Anatolia through the southern coastlands and possibly the Euphrates valley. Frequents wetlands and riverine habitats as well as woodlands and orchards and is surprisingly elusive, although its loud 'laughing' call often betrays its presence. A few pairs still breed around Dalyan (p88) and Köyceğiz Gölü in the southwest and it has been recently recorded from near Patara (Kaş) and as far north as Kuşadası. Further east the majority of the currently known Turkish population is to be found on the Tarsus river between Mersin and its outlet into the mediterranean, there being perhaps fifty pairs in this area. They have been observed occasionally at a variety of coastal locations from Antalya to Erdemli.

Pied Kingfisher An uncommon resident of coastal wetlands and major river valleys. Known from similar localities to the above species, and is also found along the Euphrates and Tigris river systems. The best site to visit for this species is Birecik/Halfeti (p30) where they are commonly seen hovering above the river. They are also found at the Seyhan/ Ceyhan delta (p37), the Göksu delta and Köyceğiz Gölü (p88) in the southwest.

Blue-cheeked Bee-eater An uncommon summer visitor to parts of the southeast where it breeds in small numbers. There are at least two breeding colonies in or near Birecik (p30), although the main one seemed to be damaged and unoccupied in 1992. Elsewhere there are a small number of records from the southeast and east but these may well relate to passage birds. There is a strong possiblity that they breed in the Arras valley on the border with Armenia. Blue-cheeked Bee-eaters prefer dry, often sandy, areas with at least some trees and usually in the vicinity of water.

Black Woodpecker A local resident of coniferous and mixed forests found in a discontinous belt from Uludağ (p85) to the Georgian/Armenian borders. As much of the forest of the Black Sea Coastlands is still little known in ornithological terms, their range can only be surmised from a scattering of records. These have mostly been from the eastern end of the Pontic Alps in the Artvin/Borçka/Arhavi area (p50) where they are probably reasonably common, and a few observations from Sumela (Trabzon), Uludağ (p85) and Kızılcahamam (p16).

White-backed Woodpecker A rare resident found in old forests containing plenty of dead wood. It has a wide range in Turkey but must be considered very rare over much of it, except perhaps for the northeast where there is likely to be a reasonable population in sub-montane pontic fir or mixed forests around Artvin, Borçka (p50) and Arhavi. Elsewhere they have been found in the western Toros (Akseki), the Bursa area near Uludağ (p85), Thrace and even at Kocaçay (p83) in the well-wooded delta.

Desert Lark A very rare resident(?) in the arid areas to the east of the Euphrates at Birecik (p30) in the southeast region. This is the only locality from which they are known, although it is quite likely that other suitable habitat along the Syrian border contains this species. They are not particularly difficult to locate among the dry hills dissected by wadis just northeast of the town, although numbers must be low as records of more than half a dozen are rare.

Yellow Wagtail The race *M. f. feldegg* is a very common summer visitor breeding in considerable numbers in many wetlands. During spring and autumn there are several other races which pass through Turkey. The huge majority of these (particularly in the east) are *M. f. beema* which breeds in Russia; these have a pale bluish grey crown, nape and ear coverts with a whitish supercilium. *M. f. thunbergi* occurs in small numbers and is distinguished by its dark grey crown, blackish ear coverts, indistinct or totally absent supercilium and yellow chin and throat. *M. f. lutea* which resembles a very yellow version of *M. f. flavissima* (yellow head with olive streaks on the crown) is a rare migrant in the east.

Citrine Wagtail A summer visitor breeding in small numbers in the east and a rare passage migrant over the rest of the country. Breeds near water in the montane steppelands around the northern and eastern shores of Van Gölü (p56) and probably further afield in the northeast. The best areas to look for them include Bendimahi (p63), Ercek Gölü (p60) and Bulanık (p75). On passage they are often found near wetlands but could turn up anywhere.

Radde's Accentor An uncommon resident or altitudinal migrant breeding in high alpine country. Inhabits ravines and boulder fields with low scrub usually above 2200m in the breeding season, and presumably moves lower in the winter months. They occur through the highest parts of the Toros in the southern coastlands, the southeast, the east and the eastern end of the Black Sea coastlands. Although nowhere common, they are somewhat unobtrusive and groups of up to forty in sheltered gullies on the slopes of Suphan Dağı (p70) in autumn indicate that they may be more common than currently thought.

Alpine Accentor A local species which can be quite common. Its preference for the highest peaks and ridges of craggy mountains can make this species difficult to find in Turkey. It has a similar range to the above species although it extends further west. The easiest sites to look for them are at Uludağ (p85) in western Anatolia and at Demirkazık (p39) in the Toros. They are one of the commonest species above 2500m in the Kackar Dagları (p49) in the eastern Pontics.

White-throated Robin A widespread but local summer visitor frequenting stony hillsides and valleys with scrub or light woodland, usually between 1000 and 2200m. Sparsely distributed in western Anatolia and the Toros mountains, this species becomes gradually more common as the anti-Toros sweeps up round in an arc to the eastern highlands. Here they are one of the more obvious species in the arid steppe characterised by *Quercus brantii*, a 'dwarf' oak which is frequent between Elazığ and Bingöl on the approaches to the eastern plateau. They are also found to the north and northeast of Van Gölü (p56). The best sites to visit when looking for this species are at Yeşilce (p34) near Gaziantep, the Kurucu Geçidi, west of Bingöl, and in the small gorges near the Şelale waterfall, northeast of Van Gölü (p65).

Pied Wheatear A rare passage migrant which possibly breeds in the extreme northeast of the country. There is a lot of confusion over the status of this species in Turkey especially with regards to identification and the apparent possibilty of hybridisation with Black-eared Wheatear in the nearest known breeding area in nearby Armenia.

Red-tailed Wheatear A rare and localised summer visitor to parts of the east and southeast. In the breeding season they inhabit barren rocky areas with boulders or scree and often with low vegetation. Regularly recorded as a breeding species from the hills to the west of Gaziantep and also on the tops of the gorge where the Euphrates flows through Halfeti (p30). Recently reported to be one of the commoner species at Nemrut Daği (p72) near Adıyaman. In the east there are records from a number of localities but apparently no regular breeding has yet been observed. They have been recorded as breeding in the Hakkari area but with few ornithologists visiting this region in the last decade there is little current information.

Graceful Warbler A local but not uncommon resident found in the southern coastlands and the southeast. Occurs as far west as Antalya but the bulk of the population is found in the Göksu delta (p37) and the Seyhan/Ceyhan delta. In the southeast more or less restricted to riverine habitat such as the reedy fringes of the Euphrates at Birecik (p30). Prefers scrubby areas with thick vegetation.

Paddyfield Warbler First discovered as a breeding species during 1986 in south Van marsh (p74) where two or three singing males were trapped and photographed. They have been noted at the same site in the breeding season in several subsequent years but have not as yet been discovered anywhere else. Otherwise a very rare passage migrant presumably somewhat overlooked because of difficulties in identifying migrant *Acrocephalus* warblers.

Upcher's Warbler A local and generally uncommon summer visitor, mainly to the southeast and to the extreme east of the southern coastlands. Breeds in rocky areas from valley levels up to 2000m often in scrubby or even cultivated areas. They are reasonably easy to locate at Yeşilce (p34) in the Kartal Dağı to the west of Gaziantep, and also in the Euphrates valley around Halfeti (p30). They have been recorded in the breeding season in the Black Sea coastlands (İspir) and more frequently in the east although it is not clear whether they breed in this region.

Olive-tree Warbler A local and uncommon summer visitor to olive groves, woodland fringes and scrub, often associated with cultivation. Recorded as a breeding species from the southern coastlands, western Anatolia, Thrace and the westernmost areas of the Black Sea coastlands. One of the best areas to look for this species is between Çanakkale and Ayvaçık. Otherwise try wooded parts of İstanbul (p15) and its environs, or around Akseki in the southern coastlands, where they have been regularly recorded.

Ménétries' Warbler An uncommon summer visitor to the southeastern region and possibly to the east. They favour scrubby areas including cultivation where there is suitable cover. They are quite common at Birecik and Halfeti (p30), and are known to occur well up the Euphrates river system, although records are few.

Rüppell's Warbler A widespread and locally common summer visitor to the southern coastlands and western Anatolia, also occurring very locally on the fringes of the central plateau, the western end of the Black Sea coastlands and the extreme southwest of the southeastern region. Breeds in areas of low scrub such as the phrygana (degraded maquis scrub) of western Anatolia and is particularly common in areas of Anatolian oak woodland.

Green Warbler A locally common summer visitor to mixed deciduous and coniferous forests in the Black Sea coastlands. They have been found in the oak/hornbeam/fir forests to the northwest of Ankara in the Abant and Yedigöller area, and in the Pontic forests of the northeast from Sumela (p47) and Sivrikaya (p43) eastwards. In the latter they can be particularly common from 500m to the tree limit. It is very likely that the wide range between these areas is occupied where suitable woodland or forest exists, the paucity of records is more likely to be from a lack of birdwatchers visiting this vast area.

Mountain Chiffchaff A locally common summer visitor(?) to the northeast and Black Sea coastlands where they occur in forest edges and scrub in the subalpine zone. Movements are not properly understood in the Turkish population, but there are a number of autumn records of birds in scrubby ravines and hillsides to the north and east of Van Gölü (p56), which resemble this species (normal Chiffchaffs migrating through this area are mostly of the greyer eastern races '*tristis*' and '*abietinus*') and perhaps indicate migrating Turkish birds. The taxonomic position of the species is currently unclear and becomes even more contentious towards the western part of the Black Sea coastlands and central plateau, where a similar Chiffchaff occurs which has sometimes been assigned a seperate race, *P. c. brevirostris*, and at the moment it is not clear whether or not Mountain Chiffchaff also occurs in this area.

Red-breasted Flycatcher A summer visitor to the Black Sea coastlands where they breed in low to moderate numbers in deciduous or mixed forests in the mountains up to 1500m. Much more widespread on passage although nowhere common. Probably breeds from the Abant Gölü area all the way along the mountains to the Georgian border where there is suitable habitat, but most breeding season records are from the northeast.

Semi-collared Flycatcher A widespread but highly localised summer visitor breeding in lightly wooded areas such as orchards, parks and large gardens. Very scarce in Thrace, parts of the east, central plateau and eastern southern coastlands, but more common in the valleys to the south of the eastern Black Sea coastlands. The Çoruh valley betwen İspir (p46) and Artvin is a very good area to look for this species as there is a good mosaic of habitats including many fruit orchards and small poplar woods, a combination not often found in eastern Turkey.

Sombre Tit A widespread but not particularly common resident. Frequents lowland plains and hilly or mountainous area with mixed woodland and scrub, particularly where 'dwarf' oaks such as *Quercus brantii* in the east and *Q. ilex* in the west are dominant. Occurs through Thrace, western Anatolia, the southern coastlands, parts of the southeast and Black Sea coastlands and the upper part of the Euphrates river system. Absent in the central plateau, the Van region and through large tracts of the other regions.

Krüper's Nuthatch A resident species which can be quite common where its prefered forest habitat is found. Shows a considerable affinity to forests or well wooded areas where the dominant species is *Pinus brutia* and its range in Turkey closely parallels the distribution of that species. It occurs in an arc through the Black Sea coastlands (less frequent in the east), through western Anatolia and south and eastwards along the Toros on the fringes of the southern coastlands and central plateau. Although they are often one of the more common species in these forests they can be unobtrusive. Good areas to look for them include the Soğuksu Milli Park near Kızılcahamam (p16), any of the forested passes on the central and western Toros and any areas of *Pinus brutia* forest above 500m in the southwest.

Eastern Rock Nuthatch A rare and local resident in the southeast and the east. Regularly recorded only from the Yeşilce area (p34) near Gaziantep and from Halfeti (p30) on the river Euphrates, but there are a number of records from other parts of the southeast which have been less frequently visited, especially in the Hakkari region (Zab Gorge, Cilo/Sat Dağı and Uludere). In the east the species has been recorded to the northeast of Van Golu and around Doğubeyazıt (p68). They prefer rocky terrain such as gorges and cliffs not dissimilar to that of its congener, Rock Nuthatch, which is also found at some of the same sites.

Wallcreeper An uncommon and local resident in high mountains and gorges with some dispersion in winter down to low altitudes. Occurs in the Toros, the Black Sea mountains, and in the east. In the highest parts of the Aladağlar massif (p39) (Toros) and the Kaçkar Dağları (p50) (eastern Pontics) they can be quite common.

Masked Shrike A reasonably common summer visitor to western Anatolia, the southern coastlands and parts of the southeast and east. Their habitat preferences follow that of Rüppell's Warbler and Sombre Tit although they are more frequent in lowland cultivation with orchards and olive groves. It is not known whether the population in the arid steppe and *Quercus brantii* zone centred between Bingöl and Elazığ is isolated from, or contiguous with, the rest of the Turkish distribution.

Rose-coloured Starling The status of this species in Turkey is uncertain. It seems to be an invasive nomadic species mostly to the eastern two-thirds of the country although in some years it does reach the western coasts. Breeding has been proven in the east, and the large numbers of this species seen throughout the summer months in many years must mean that there is a possibilty that they may breed on occasions.

Dead Sea Sparrow Resident in a very few localities but can be quite abundant in these and is perhaps expanding its range. Known to breed along the Euphrates at Birecik and Halfeti (p30) and has been recorded much further up the river. Also found in adjacent parts of the southern coastlands notably at the Göksu delta (p37) and the Seyhan/Ceyhan delta. Frequents scrub, orchards and particularly tamarisk, usually near water.

Pale Rock Sparrow A rather enigmatic species. Apparently a late arriving summer visitor in variable numbers to the southeast and immediately adjacent parts of the southern coastlands and the east. May well be irruptive and/or nomadic and numbers vary considerably from year to year at its few known breeding sites. Frequents arid or semi-arid country often in rocky terrain, but also appears in grassland and cultivation. In good years this species can be found in the region bordering Syria from west of Gaziantep eastwards to the Hakkari area. The Yeşilce area (p34) is a good area to look for them, as is Halfeti (p30) and further to the east around İdil and Cizre (p35). Has been recorded in the mountains on the south shore of Van Gölü (p74) in the breeding season.

Yellow-throated Sparrow A rare summer visitor to the southeastern region, currently known to breed in the environs of the Euphrates river at Birecik and Halfeti (p30), and also from the Tigris valley near Cizre (p35). Inhabits woodland and orchards, cultivated areas, and villages where there are some trees. In Turkey seems to show a marked preference for pistachio orchards and, as there is a vast acreage of these trees in the southeast, they may well have been significantly under-recorded. This species is quite an early autumn migrant, often having left its breeding grounds by late August.

Red-fronted Serin A locally common resident in the subalpine zones of most Turkish mountain areas. Usually found between 1800m and 3000m where there is a mix of alpine meadows, sparse woodland and scrub, especially juniper. Found in the eastern half of the Black Sea coastlands, the Toros range in the southern coastlands, in parts of the east and with an isolated outpost at Uludağ (p85) in western Anatolia. They form flocks in late summer and autumn and groups of a couple of hundred birds can sometimes be found in suitable areas such as at Demirkazık (p39) in the Aladağlar.

Crimson-winged Finch A moderately common resident and partial migrant of rocky mountainous areas, usually above 2000m in the breeding season. Distributed patchily throughout the east, eastern Black Sea coastlands, the southeast, isolated volcanic areas on the fringes of the central plateau, and the eastern parts of the Toros range. The volcanic areas around Van Gölü are good areas to look for this species such as Nemrut Dağı (p72) and in particular the lavafields on the slopes of Tendürek Dağı on the road between Çaldıran (p65) and Doğubeyazıt (p68).

Desert Finch An uncommon and local resident of pistachio orchards, olive groves and other cultivation in arid areas. Found only in the southeast and immediatly adjacent areas of the southern coastlands. Regularly recorded from the Kartal Dağı near Gaziantep at Yeşilce (p34) and in the area between Birecik and Halfeti (p30). There are occasional observations further up the Euphrates river system e.g. at Adıyaman, and as with Yellow-throated Sparrow, this species is likely to be more common than records suggest.

Mongolian Trumpeter Finch Current status uncertain. Only recently discovered in eastern Turkey, this species is presumably a resident or partial migrant which has probably expanded into the country in the last few years. The first breeding record for Turkey was in 1990 at Asagı Mutlu (north of Çaldıran). Breeding probably occurred during 1992 at Doğubeyazıt (p68), in the hills immediately east of the İşak Paş a, and in the same year there have been a number of sightings around Ararat and İğdir.

Trumpeter Finch Probably a very rare and irregular summer visitor to the southeast, where breeding has occurred on at least one occasion. Most records come from the Euphrates valley and the hills west of Gaziantep. Prefers bare rocky hill-sides and wadis in stony deserts or semi-desert.

Cinereous Bunting A very local summer visitor to dry rocky slopes with sparse vegetation found from low altitudes up to the tree-limit, although usually at medium altitudes. The race *cineracea* is known from western Anatolia, and the fringes of the southern coastlands and the central plateau. The race *semenowi* is found in the southeast, and in the extreme eastern end of the southern coastlands. Also found in the east around the southern shore of Van Gölü (p74) and possibly elsewhere in the region. In all of these areas it is scarce and is known to breed in only a handful of sites in each region. The Kartal Dağı west of Gaziantep at Yeşilce (p34) and the gorge tops where the Euphrates flows past Halfeti (p30) are good sites to look for this species.

Grey-necked Bunting A local and uncommon summer visitor to dry and barren mountainsides with sparse vegetation, usually above 2000m. Breeds in the east and adjacent parts of the southeast from Ağrı Daği southwards to Hakkari and Yüksekova. Has also occasionally been recorded from the Aladağlari (p39) in the eastern Toros during the breeding season. A very unobtrusive species which is probably more common within its range than previously realised. The hills to the east of Van (p56) and the very barren mountains south of Erçek Gölü (p60) are particularly good areas for this species. In autumn, flocks of over forty birds have been observed quite close to the lake level on the southern shores of Van Gölü (p74).

FULL SPECIES LIST

The taxonomic sequence of this list follows Voous (1977). Vagrant records are as reported and are not necessarily confirmed.

Key to checklist

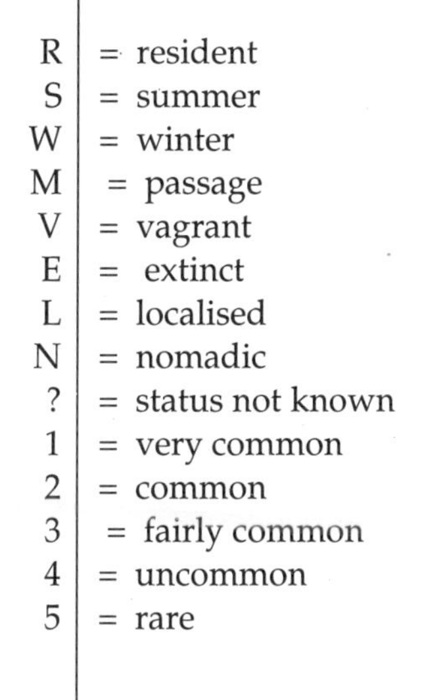

R	= resident
S	= summer
W	= winter
M	= passage
V	= vagrant
E	= extinct
L	= localised
N	= nomadic
?	= status not known
1	= very common
2	= common
3	= fairly common
4	= uncommon
5	= rare

5W ☐	Red-throated Diver (Gavia stellata)
5SL? ☐	Black-throated Diver (Gavia arctica)
5W ☐	Great Northern Diver (Gavia immer)
2R ☐	Little Grebe (Tachybaptus ruficollis)
2RL ☐	Great Crested Grebe (Podiceps cristatus)
4SL3M5W ☐	Red-necked Grebe (Podiceps grisegena)
5W ☐	Slavonian Grebe (Podiceps auritus)
2SL1ML ☐	Black-necked Grebe (Podiceps nigricollis)
2S ☐	Cory's Shearwater (Calonectris diomedea)
3SM ☐	Manx Shearwater (Puffinus puffinus)
5SM ☐	Storm Petrel (Hydrobates pelagicus)
2RL ☐	Cormorant (Phalacrocorax carbo)
4RL ☐	Shag (Phalacrocorax aristotelis)
2RL ☐	Pygmy Cormorant (Phalacrocorax pygmeus)
E ☐	Darter (Anhinga rufa)
4SML(N)SW ☐	White Pelican (Pelecanus onocrotalus)
4RL(N) ☐	Dalmatian Pelican (Pelecanus crispus)
5RL ☐	Bittern (Botaurus stellaris)
1SM3W ☐	Little Bittern (Ixobrychus minutus)
3SML ☐	Night Heron (Nycticorax nycticorax)
2SM ☐	Squacco Heron (Ardeola ralloides)
5SML ☐	Cattle Egret (Bulbulcus ibis)
1SM3W ☐	Little Egret (Egretta garzetta)
3SM ☐	Great White Egret (Egretta alba)
2R ☐	Grey Heron (Ardea cinerea)
2SM ☐	Purple Heron (Ardea purpurea)
V ☐	Yellow-billed Stork (Mycteria ibis)

3SML ☐ Black Stork (Ciconia nigra)
1SM5W ☐ White Stork (Ciconia ciconia)
4SML ☐ Glossy Ibis (Plegadis falcinellus)
E? ☐ Bald Ibis (Geronticus eremita)
3SML5W ☐ Spoonbill (Platalea leucorodia)
2RL(N) ☐ Greater Flamingo (Phoenicopterus ruber)
5R4W ☐ Mute Swan (Cygnus olor)
5W ☐ Bewick's Swan (Cygnus columbianus)
5W ☐ Whooper Swan (Cygnus cygnus)
5W ☐ Bean Goose (Anser fabalis)
2WML ☐ White-fronted Goose (Anser albifrons)
V ☐ Lesser White-fronted Goose (Anser erythropus)
2RL ☐ Greylag Goose (Anser anser)
5W ☐ Red-breasted Goose (Branta ruficollis)
1RL ☐ Ruddy Shelduck (Tadorna ferruginea)
3R ☐ Shelduck (Tadorna tadorna)
1W ☐ Wigeon (Anas penelope)
V? ☐ Falcated Duck (Anas falcata)
2R ☐ Gadwall (Anas strepera)
5SL1WM ☐ Teal (Anas crecca)
1R ☐ Mallard (Anas platyrhynchos)
5SL2WML ☐ Pintail (Anas acuta)
3SL1MW? ☐ Garganey (Anas querquedula)
4SL1MW ☐ Shoveler (Anas clypeata)
4SL5W ☐ Marbled Duck (Marmaronetta angustirostris)
3SL2WM ☐ Red-crested Pochard (Netta rufina)
4SL1MW ☐ Pochard (Aythya ferina)
4RL ☐ Ferruginous Duck (Aythya nyroca)
5SL2WM ☐ Tufted Duck (Aythya fuligula)
4W ☐ Scaup (Aythya marila)
5SLW? ☐ Long-tailed Duck (Clangula hyemalis)
V ☐ Common Scoter (Melanitta nigra)
5SL4W ☐ Velvet Scoter (Melanitta fusca)
5SL?4W ☐ Goldeneye (Bucephala clangula)
4W ☐ Smew (Mergus albellus)
4W ☐ Red-breasted Merganser (Mergus serrator)
5W ☐ Goosander (Mergus merganser)
4SL3ML2WL ☐ White-headed Duck (Oxyura leucocephala)
4SL1M ☐ Honey Buzzard (Pernis apivorus)
V ☐ Black-shouldered Kite (Elanus caeruleus)
2RL ☐ Black Kite (Milvus migrans)
5M?W ☐ Red Kite (Milvus milvus)
5RL ☐ White-tailed Eagle (Haliaeetus albicilla)
4R ☐ Lammergeier (Gypaetus barbatus)
3S ☐ Egyptian Vulture (Neophron percnopterus)
3R ☐ Griffon Vulture (Gyps fulvus)
4R ☐ Black Vulture (Aegypius monachus)
2S ☐ Short-toed Eagle (Circaetus gallicus)
2R ☐ Marsh Harrier (Circus aeruginosus)

4WM ☐ Hen Harrier (Circus cyaneus)
4M ☐ Pallid Harrier (Circus macrourus)
4SL2M ☐ Montagu's Harrier (Circus pygargus)
4S3WM ☐ Goshawk (Accipiter gentilis)
2SMW ☐ Sparrowhawk (Accipiter nisus)
5SL3M ☐ Levant Sparrowhawk (Accipiter brevipes)
1SMW? ☐ Buzzard (Buteo buteo)
2R ☐ Long-legged Buzzard (Buteo rufinus)
5W ☐ Rough-legged Buzzard (Buteo lagopus)
3SL2M ☐ Lesser Spotted Eagle (Aquila pomarina)
5SL4MW ☐ Spotted Eagle (Aquila clanga)
4M ☐ Steppe Eagle (Aquila rapax)
4RM ☐ Imperial Eagle (Aquila heliaca)
3R ☐ Golden Eagle (Aquila chrysaetos)
3SLM ☐ Booted Eagle (Hieraaetus pennatus)
5RL? ☐ Bonelli's Eagle (Hieraaetus fasciatus)
5S4M ☐ Osprey (Pandion haliaetus)
2SL5W ☐ Lesser Kestrel (Falco naumanni)
1R ☐ Kestrel (Falco tinnunculus)
2M5SL? ☐ Red-footed Falcon (Falco vespertinus)
4MW ☐ Merlin (Falco columbarius)
2SM ☐ Hobby (Falco subbuteo)
4SM ☐ Eleonora's Falcon (Falco eleonorae)
4R? ☐ Lanner (Falco biarmicus)
4SM? ☐ Saker (Falco cherrug)
3RM ☐ Peregrine (Falco peregrinus)
5RL ☐ Caucasian Black Grouse (Tetrao mlokosiewiczi)
3RL ☐ Caspian Snowcock (Tetraogallus caspius)
1R ☐ Chukar (Alectoris chukar)
5RL ☐ See-see (Ammoperdix griseogularis)
4RL ☐ Black Francolin (Francolinus francolinus)
3R? ☐ Grey Partridge (Perdix perdix)
1SM ☐ Quail (Coturnix coturnix)
4R? ☐ Pheasant (Phasianus colchicus)
2R ☐ Water Rail (Rallus aquaticus)
4M?S ☐ Spotted Crake (Porzana porzana)
4M?S ☐ Little Crake (Porzana parva)
4M?S ☐ Baillon's Crake (Porzana pusilla)
5M ☐ Corncrake (Crex crex)
1R ☐ Moorhen (Gallinula chloropus)
5RL ☐ Purple Gallinule (Porphyrio porphyrio)
S1MW ☐ Coot (Fulica atra)
4SL3M ☐ Crane (Grus grus)
5SML ☐ Demoiselle Crane (Anthropoides virgo)
5RS?W ☐ Little Bustard (Tetrax tetrax)
4RL ☐ Great Bustard (Otis tarda)
3SL ☐ Oystercatcher (Haematopus ostralegus)
1SM2W ☐ Black-winged Stilt (Himantopus himantopus)

1SM2W ☐	Avocet (Recurvirostra avosetta)
V ☐	Crab Plover (Dromas ardeola)
4SM ☐	Stone-curlew (Burhinus oedicnemus)
5SL? ☐	Cream-coloured Courser (Cursorius cursor)
3SL2M ☐	Collared Pratincole (Glareola pratincola)
4MS? ☐	Black-winged Pratincole (Glareola nordmanni)
2SM4W ☐	Little Ringed Plover (Charadrius dubius)
4WM ☐	Ringed Plover (Charadrius hiaticula)
2SML3W ☐	Kentish Plover (Charadrius alexandrinus)
V ☐	Lesser Sand Plover (Charadrius mongolus)
4SML ☐	Greater Sand Plover (Charadrius leschenaultii)
5M ☐	Caspian Plover (Charadrius asiaticus)
5M ☐	Dotterel (Charadrius morinellus)
V ☐	Pacific Golden Plover (Pluvialis fulva)
3W? ☐	Golden Plover (Pluvialis apricaria)
4WM ☐	Grey Plover (Pluvialis squatarola)
2SML ☐	Spur-winged Plover (Hoplopterus spinosus)
5R? ☐	Red-wattled Plover (Hoplopterus indicus)
5M ☐	Sociable Plover (Chettusia gregaria)
5SML ☐	White-tailed Plover (Chettusia leucura)
1R ☐	Lapwing (Vanellus vanellus)
V ☐	Knot (Charadrius canutus)
4WM ☐	Sanderling (Charadrius alba)
1M4W ☐	Little Stint (Charadrius minuta)
4M ☐	Temminck's Stint (Charadrius temminckii)
2M ☐	Curlew Sandpiper (Charadrius ferruginea)
3WM ☐	Dunlin (Charadrius alpina)
3ML ☐	Broad-billed Sandpiper (Limicola falcinellus)
1M3W ☐	Ruff (Philomachus pugnax)
4WM ☐	Jack Snipe (Lymnocryptes minimus)
2WM ☐	Snipe (Gallinago gallinago)
4M ☐	Great Snipe (Gallinago media)
4MW ☐	Woodcock (Scolopax rusticola)
3M4W ☐	Black-tailed Godwit (Limosa limosa)
5MW ☐	Bar-tailed Godwit (Limosa lapponica)
4M ☐	Whimbrel (Numenius phaeopus)
5M ☐	Slender-billed Curlew (Numenius tenuirostris)
4MW ☐	Curlew (Numenius arquata)
3M4W ☐	Spotted Redshank (Tringa erythropus)
1SMW ☐	Redshank (Tringa totanus)
4M ☐	Marsh Sandpiper (Tringa stagnatilis)
2M3W ☐	Greenshank (Tringa nebularia)
2MW ☐	Green Sandpiper (Tringa ochropus)
1M5W ☐	Wood Sandpiper (Tringa glareola)
4M ☐	Terek Sandpiper (Xenus cinereus)
3SL1MSW ☐	Common Sandpiper (Actitis hypoleucos)
4M ☐	Turnstone (Arenaria interpres)
4ML ☐	Red-necked Phalarope (Phalaropus lobatus)

V ☐ Pomarine Skua (Stercorarius pomarinus)
V ☐ Arctic Skua (Stercorarius parasiticus)
V ☐ Long-tailed Skua (Stercorarius longicaudus)
V ☐ Great Skua (Stercorarius skua)
V ☐ Great Black-headed Gull (Larus icthyaetus)
3SL2MSW ☐ Mediterranean Gull (Larus melanocephalus)
3MW ☐ Little Gull (Larus minutus)
1R ☐ Black-headed Gull (Larus ridibundus)
3SM ☐ Slender-billed Gull (Larus genei)
4R ☐ Audouin's Gull (Larus audouinii)
4MW ☐ Common Gull (Larus canus)
3MW ☐ Lesser Black-backed GULL (Larus fuscus)
1R ☐ Herring Gull (Larus argentatus)
V ☐ Great Black-backed Gull (Larus marinus)
V ☐ Kittiwake (Rissa tridactyla)
2SM ☐ Gull-billed Tern (Gelochelidon nilotica)
4SL3MSW ☐ Caspian Tern (Sterna caspia)
V ☐ Lesser Crested Tern (Sterna benegalensis)
4M3W ☐ Sandwich Tern (Sterna sandvicensis)
1SM ☐ Common Tern (Sterna hirundo)
V ☐ Arctic Tern (Sterna paradisaea)
3SL2M ☐ Little Tern (Sterna albifrons)
4SL?2MSW ☐ Whiskered Tern (Chlidonias hybridus)
5SL?3M ☐ Black Tern (Chlidonias niger
5SL?1M ☐ White-winged Black Tern (Chlidonias leucopterus)
V ☐ Spotted Sandgrouse (Pterocles senegallus)
3R ☐ Black-bellied Sandgrouse (Pterocles orientalis)
4RL(N) ☐ Pin-Tailed Sandgrouse (Pterocles alchata)
1R ☐ Rock Dove (Columba livia)
3R2M ☐ Stock Dove (Columba oenas)
1R ☐ Woodpigeon (Columba palumbus)
1R ☐ Collared Dove (Streptopelia decaocto)
1SM ☐ Turtle Dove (Streptopelia turtur)
3RL ☐ Laughing Dove (Streptopelia senegalensis)
5SL? ☐ Great Spotted Cuckoo (Clamator glandarius)
1S ☐ Cuckoo (Cuculus canorus)
4RL ☐ Barn Owl (Tyto alba)
5SL ☐ Striated Scops Owl (Otus brucei)
2S ☐ Scops Owl (Otus scops)
4RL ☐ Eagle Owl (Bubo bubo)
5RL ☐ Brown Fish Owl (Ketupa zeylonensis)
1R ☐ Little Owl (Athene noctua)
3RL ☐ Tawny Owl (Strix aluco)
5SL?4W ☐ Long-eared Owl (Asio otus)
5SL4W ☐ Short-eared Owl (Asio flammeus)
5RL? ☐ Tengmalm's Owl (Aegolius funereus)
3S2M ☐ Nightjar (Caprimulgus europaeus)
1S ☐ Swift (Apus apus)

5SL ☐	Pallid Swift (Apus pallidus)
2SM ☐	Alpine Swift (Apus melba)
5SL ☐	Little Swift (Apus affinis)
5RL ☐	White-breasted Kingfisher (Halcyon smyrnensis)
4SL2WM ☐	Kingfisher (Alcedo atthis)
4RL ☐	Pied Kingfisher (Ceryle rudis)
5SL4M ☐	Blue-cheeked Bee-eater (Merops superciliosus)
1SM ☐	Bee-eater (Merops apiaster)
1SM ☐	Roller (Coracias garrulus)
2SM ☐	Hoopoe (Upupa epops)
4SL3MSW ☐	Wryneck (Jynx torquilla)
5SL? ☐	Grey-headed Woodpecker (Picus canus)
2R ☐	Green Woodpecker (Picus viridis)
4RL ☐	Black Woodpecker (Dryocopus martius)
2R ☐	Great Spotted Woodpecker (Dendrocopos major)
2R ☐	Syrian Woodpecker (Dendrocopos syriacus)
3RL ☐	Middle Spotted Woodpecker (Dendrocopos medius)
4RL ☐	White-backed Woodpecker (Dendrocopos leucotos)
3R ☐	Lesser Spotted Woodpecker (Dendrocopos minor)
5R ☐	Desert Lark (Ammomanes deserti)
2R ☐	Calandra Lark (Melanocorypha calandra)
3S ☐	Bimaculated Lark (Melanocorypha bimaculata)
V ☐	White-winged Lark (Melanocorypha leucoptera)
V ☐	Black Lark (Melanocorypha yeltoniensis)
2S ☐	Short-toed Lark (Calandrella brachydactyla)
2SL ☐	Lesser Short-toed Lark (Calandrella rufescens)
1R ☐	Crested Lark (Galerida cristata)
3SM ☐	Woodlark (Lullula arborea)
2R ☐	Skylark (Alauda arvensis)
2RL ☐	Shore Lark (Eremophila alpestris)
1SM ☐	Sand Martin (Riparia riparia)
1SMR ☐	Crag Martin (Ptyonoprogne rupestris)
1SM ☐	Swallow (Hirundo rustica)
2SML ☐	Red-rumped Swallow (Hirundo daurica)
1SM ☐	House Martin (Delichon urbica)
V ☐	Richard's Pipit (Anthus novaeseelandiae)
1SM ☐	Tawny Pipit (Anthus campestris)
3S1M ☐	Tree Pipit (Anthus trivialis)
2MW ☐	Meadow Pipit (Anthus pratensis)
2ML ☐	Red-throated Pipit (Anthus cervinus)
2RL ☐	Water Pipit (Anthus spinoletta)
1SM ☐	Yellow Wagtail (Motacilla flava feldegg)
1M ☐	(Motacilla flava beema)
4M ☐	(Motacilla flava thunbergi)
5M ☐	(Motacilla flava lutea)
4SML ☐	Citrine Wagtail (Motacilla citreola)
2SL1WM ☐	Grey Wagtail (Motacilla cinerea)
2S1MW ☐	Pied Wagtail (Motacilla alba)

4RL ☐ Yellow-vented Bulbul (Pycnonotus xanthopygos)
5W ☐ Waxwing (Bombycilla garrulus)
2RL ☐ Dipper (Cinclus cinclus)
2R1W ☐ Wren (Troglodytes troglodytes)
3RL ☐ Dunnock (Prunella modularis)
4RL ☐ Radde's Accentor (Prunella ocularis)
3RL ☐ Alpine Accentor (Prunella collaris)
2SM ☐ Rufous Bush-robin (Cercotricas galactotes)
3R2W ☐ Robin (Erithacus rubecula)
4M ☐ Thrush Nightingale (Luscinia luscinia)
2S1M ☐ Nightingale (Luscinia megarhynchos)
5SL4M ☐ Bluethroat (Luscinia svecica)
2SLM ☐ White-throated Robin (Irania gutturalis)
1SM ☐ Black Redstart (Phoenicurus ochruros)
2SLM ☐ Redstart (Phoenicurus phoenicurus)
3SL1M ☐ Whinchat (Saxicola rubetra)
1R ☐ Stonechat (Saxicola torquata)
1SM ☐ Isabelline Wheatear (Oenanthe isabellina)
1SM ☐ Wheatear (Oenanthe oenanthe)
5SML ☐ Pied Wheatear (Oenanthe pleschanka)
V ☐ Cyprus Pied Wheatear (Oenanthe cypriaca)
1SM ☐ Black-eared Wheatear (Oenanthe hispanica)
V ☐ Desert Wheatear (Oenanthe deserti)
3SMW? ☐ Finsch's Wheatear (Oenanthe finschii)
V ☐ Red-rumped Wheatear (Oenanthe moesta)
4SLM ☐ Red-tailed Wheatear (Oenanthe xanthoprymna)
V ☐ White-crowned Black Wheatear (Oenanthe leucopyga)
2SLM ☐ Rock Thrush (Monticola saxatilis)
2SMR ☐ Blue Rock Thrush (Monticola solitarius)
4SML ☐ Ring Ouzel (Turdus torquatus)
1R ☐ Blackbird (Turdus merula)
2W ☐ Fieldfare (Turdus pilaris)
3RL ☐ Song Thrush (Turdus philomelos)
2W ☐ Redwing (Turdus iliacus)
2R ☐ Mistle Thrush (Turdus viscivorus)
1R ☐ Cetti's Warbler (Cettia cetti)
4R ☐ Fan-tailed Warbler (Cisticola juncidis)
4R ☐ Graceful Warbler (Prinia gracilis)
5M ☐ Grasshopper Warbler (Locustella naevia)
5M ☐ River Warbler (Locustella fluviatilis)
4SML ☐ Savi's Warbler (Locustella luscinoides)
3SML4WL ☐ Moustached Warbler (Acrocephalus melanopogon)
V ☐ Aquatic Warbler (Acrocephalus paludicola)
3SML ☐ Sedge Warbler (Acrocephalus schoenobaenus)
5SL4M ☐ Paddyfield Warbler (Acrocephalus agricola)
4SL2M ☐ Marsh Warbler (Acrocephalus palustris)
3S1M ☐ Reed Warbler (Acrocephalus scirpaceus)
1SM ☐ Great Reed Warbler (Acrocephalus arundinaceus)

1SM ☐ Olivaceous Warbler (Hippolais pallida)
5M ☐ Booted Warbler (Hippolais caligata)
4SML ☐ Upcher's Warbler (Hippolais languida)
4SML ☐ Olive-tree Warbler (Hippolais olivetorum)
5SL?4M ☐ Icterine Warbler (Hippolais icterina)
V ☐ Melodious Warbler (Hippolais polyglotta)
V ☐ Spectacled Warbler (Sylvia conspicillata)
3SML ☐ Subalpine Warbler (Sylvia cantillans)
4SML ☐ Ménétries Warbler (Sylvia mystacea)
2RL ☐ Sardinian Warbler (Sylvia melanocephala)
V ☐ Cyprus Warbler (Sylvia melanothorax)
2SM ☐ Rüppell's Warbler (Sylvia rueppelli)
V ☐ Desert Warbler (Sylvia nana)
3SML ☐ Orphean Warbler (Sylvia hortensis)
4SL?3M ☐ Barred Warbler (Sylvia nisoria)
1SM ☐ Lesser Whitethroat (Sylvia curruca)
1SM ☐ Whitethroat (Sylvia communis)
4SL2M ☐ Garden Warbler (Sylvia borin)
3SL1M ☐ Blackcap (Sylvia atricapilla)
2SML ☐ Green Warbler (Phylloscopus nitidus)
V ☐ Greenish Warbler (Phylloscopus trochiloides)
V ☐ Yellow-browed Warbler (Phylloscopus inornatus)
3S2M ☐ Bonelli's Warbler (Phylloscopus bonelli)
2M ☐ Wood Warbler (Phylloscopus sibilatrix)
3SML ☐ Mountain Chiffchaff (Phylloscopus sindianus)
3S1M ☐ Chiffchaff (Phylloscopus collybita)
2M ☐ Willow Warbler (Phylloscopus trochilus)
2R ☐ Goldcrest (Regulus regulus)
3RL ☐ Firecrest (Regulus ignicapillus)
4S1M ☐ Spotted Flycatcher (Muscicapa striata)
4S3M ☐ Red-Breasted Flycatcher (Ficedula parva)
4SL3M ☐ Semi-collared Flycatcher (Ficedula semitorquata)
2M ☐ Collared Flycatcher (Ficedula albicollis)
3M ☐ Pied Flycatcher (Ficedula hypoleuca)
3R(N) ☐ Bearded Tit (Panurus biarmicus)
2R ☐ Long-tailed Tit (Aegithalos caudatus)
3RL ☐ Marsh Tit (Parus palustris)
2R ☐ Sombre Tit (Parus lugubris)
1R ☐ Coal Tit (Parus ater)
1R ☐ Blue Tit (Parus caeruleus)
1R ☐ Great Tit (Parus major)
2R ☐ Krüper's Nuthatch (Sitta krueperi)
2R ☐ Nuthatch (Sitta europaea)
4RL ☐ Great Rock Nuthatch (Sitta tephronota)
1R ☐ Rock Nuthatch (Sitta neumayer)
3RL ☐ Wallcreeper (Tichodroma muraria)
3R ☐ Treecreeper (Certhia familiaris)
3R ☐ Short-toed Treecreeper (Certhia brachydactyla)

2RL ☐ Penduline Tit (Remiz pendulinus)
1SM ☐ Golden Oriole (Oriolus oriolus)
V ☐ Isabelline Shrike (Lanius isabellinus)
1SM ☐ Red-backed Shrike (Lanius collurio)
3SM ☐ Lesser Grey Shrike (Lanius minor)
5W ☐ Great Grey Shrike (Lanius excubitor)
2SM ☐ Woodchat Shrike (Lanius senator)
2SM ☐ Masked Shrike (Lanius nubicus)
1R ☐ Jay (Garrulus glandarius)
1R ☐ Magpie (Pica pica)
V ☐ Nutcracker (Nucifraga caryocatactes)
3RL ☐ Alpine Chough (Pyrrhocorax graculus)
3RL ☐ Chough (Pyrrhocorax pyrrhocorax)
1R ☐ Jackdaw (Corvus monedula)
2RL ☐ Rook (Corvus frugilegus)
1R ☐ Carrion (Hooded) Crow (Corvus corone)
V ☐ Brown-necked Raven (Corvus ruficollis)
2R ☐ Raven (Corvus corax)
1R ☐ Starling (Sturnus vulgaris)
2SML(N) ☐ Rose-coloured Starling (Sturnus roseus)
1R ☐ House Sparrow (Passer domesticus)
2RSL? ☐ Spanish Sparrow (Passer hispaniolensis)
5SL ☐ Dead Sea Sparrow (Passer moabiticus)
2R ☐ Tree Sparrow (Passer montanus)
4SL(N) ☐ Pale Rock Sparrow (Petronia brachydactyla)
5SL ☐ Yellow-throated Sparrow (Petronia xanthocollis)
1R ☐ Rock Sparrow (Petronia petronia)
3RL ☐ Snow Finch (Montifringilla nivalis)
1R ☐ Chaffinch (Fringilla coelebs)
3WM ☐ Brambling (Fringilla montifringilla)
2RL ☐ Red-fronted Serin (Serinus pusillus)
2R ☐ Serin (Serinus serinus)
2R ☐ Greenfinch (Carduelis chloris)
1R ☐ Goldfinch (Carduelis carduelis)
4RL3WM ☐ Siskin (Carduelis spinus)
1R ☐ Linnet (Carduelis cannabina)
3RL ☐ Twite (Carduelis flavirostris)
V ☐ Redpoll (Carduelis flammea)
3R ☐ Crossbill (Loxia curvirostra)
3R ☐ Crimson-winged Finch (Rhodopechys sanguinea)
4R ☐ Desert Finch (Rhodospiza obsoleta)
5SL ☐ Mongolian Trumpeter Finch (Bucanetes mongolicus)
5SMLN? ☐ Trumpeter Finch (Bucanetes githagineus)
2S ☐ Scarlet Rosefinch (Carpodacus erythrinus)
V ☐ Great Rosefinch (Carpodacus rubicilla)
3R ☐ Bullfinch (Pyrrhula pyrrhula)
4R3M ☐ Hawfinch (Coccothraustes coccothraustes)
V ☐ Snow Bunting (Plectrophenax nivalis)

V ☐ Pine Bunting (Emberiza leucocephalos)
5SL?4MW ☐ Yellowhammer (Emberiza citrinella)
3R ☐ Cirl Bunting (Emberiza cirlus)
2R ☐ Rock Bunting (Emberiza cia)
4SML ☐ Cinereous Bunting (Emberiza cineracea)
3S1M ☐ Ortolan Bunting (Emberiza hortulana)
4SML ☐ Grey-Necked Bunting (Emberiza buchanani)
2SM ☐ Cretzschmar's Bunting (Emberiza caesia)
V ☐ Rustic Bunting (Emberiza rustica)
V ☐ Little Bunting (Emberiza pusilla)
V ☐ Yellow-breasted Bunting (Emberiza aureola)
4RL ☐ Reed Bunting (Emberiza schoeniclus)
V ☐ Red-headed Bunting (Emberiza bruniceps)
1SM ☐ Black-headed Bunting (Emberiza melanocephala)
1R ☐ Corn Bunting (Milaria calandra)

CONSERVATION IN TURKEY

Turkey does not have a long history of conservation, although it has come to the fore on both the public and private agenda in recent years. Turkey is a signatory to the World Heritage Convention (it has designated one World Heritage Site), the Bern Convention and the Barcelona Convention, but it has, like many other countries, been slow to implement them. Turkey has a total of 19 National Parks (1989) which are largely concentrated in the central and western regions. This reflects the bias in management towards toruism and recreation. A number of large National Parks are currently being designated in the east of the country such as the Kaçkar mountains around Ayder, and Nemrut Daği near Bitlis. Turkey has its own very active conservation organisation, the DHKD (Doğal Hayatı Koruma Derneği) which from a small but influential membership has done a lot to further the cause of conservation in recent years. As with anywhere, gaining the support of local people gets results and the designation of the Dalyan area as a Specially Protected Environmental Area, and the local Council ban on hunting White-headed Duck on Burdur Gölü show what can be done (sadly both these may shortly fall by the wayside). Outside of Turkey the Ornithological Society of the Middle East (OSME) is a major forum for discussion, collation of ornithological data and encouraging research and conservation within Turkey and they now work very closely with the DHKD. In short, your observations, whether a species list or a detailed account of a site visited, are valuable and can be very useful in aiding the conservation of species and areas. We would like to encourage all birdwatchers visiting Turkey to provide details of their sightings and in particular on any material changes to the sites they have visited to the above two organisations. The addresses are below:

Ornithological Society of the Middle East (OSME)
c/o The Lodge, Sandy, Bedfordshire, SG19 2DL, UK

Doğal Hayatı Koruma Derneği (DHKD)
PK 18, 80810 Bebek, Istanbul, Turkey.

FURTHER READING

The diversity of wildlife in a large country such as Turkey makes it impractical to include species checklists other than for birds, the main subject matter, in a guide such as this. A general guide to the better fieldguides and books for other groups is included in this section in place of the checklists.

Turkey is home to around ten thousand species of higher plants and thus the limitations of any fieldguide to the flowers of Turkey are huge. In fact one of the best published floras of any country or region is the 'Flora of Turkey', edited by P. H. Davis, and published by Edinburgh University Press. The first volume came out in the sixties and this monumental work was eventually completed in the early eighties, and is indeed a lifetime's work. The ten volumes cost around £65 each. This is a highly technical flora and would only interest the serious botanist! For those interested in at least being able to place plants in a family, or identifying the commoner and more spectacular species there are a number of useful guides. Oleg Polunin's 'Flowers of Greece and the Balkans' will help you to do this, especially in the west of the country, and the newly published 'Mediterranean Wildflowers' by Marjorie Blamey and Christopher Grey-Wilson is also quite good. But don't expect too much - a lot of plant families in Turkey are representatives of more eastern floral elements and thus there are yawning gaps in these books. Polunin's guide is also out of print and practically impossible to get hold off, hopefully a new printing is on the way! For particular groups of plants there are a number of specialist monographs available - Summerfield Books in Brough near Kirby Stephen in Cumbria (Tel: 017683 41577) are specialist dealers in both new and second-hand botanical books.

Orchids are one group that is well covered in Turkey, most of new European Orchid fieldguides actually cover all the species in the country, even to the Iranian and Armenian borders. The best is Karl Peter Buttler's 'Orchids of Britain and Europe', published by Crowood Press. This is a photographic guide, a medium which works much better with this family than other plants, and the text detail is very good.

Butterflies are adequately covered by the Collins fieldguide to the Butterflies of Britain and Europe by Higgins and Hargreaves. Unfortunately this guide maps species for the whole of Turkey, although the fieldguide is only designed to go as far as Greece, with the result that the distribution maps are rather vague for Turkey. Also, many species occuring in Turkey are not covered by this book. This is much more of a problem in the east of the country, but even on the west coast we have found species not covered by this book.

Reptiles and amphibians are well covered by the Collins European fieldguide as are the mammals should you be lucky enough to see many of them!

For those interested in learning a little of the language there are a number of phrasebooks on the market. One of the easiest to follow is

'Quick & Easy Turkish' by Necdet Teymur, a Langenscheidt Teach Yourself phrasebook published by Hodder and Stoughton. Much harder, but going a lot further, is 'Colloquial Turkish' by Yusuf Mardin. This is a real slog and only for those really interested in being able to follow conversations on the current political climate and exactly which valley that Blackcock was seen feeding in the Rhododendrons!

Journals and magazines: Sandgrouse, the now twice yearly journal of the Ornithological Society of the Middle East and their regular bulletins are excellent for keeping up to date with the latest sightings in Turkey. The journal itself keeps abreast of the latest research work on the regions avifauna and on conservation issues. The mainstream birdwatching press sometimes have articles on Turkey, and the Western Paleartic News section in the monthly Birding World magazine usually has an update on the more unusual (unconfirmed) avian records from Turkey. Those interested in the flora might like to take a look at the Alpine Garden Society's quarterly journal - Turkey features quite frequently in its pages.